UNECE

Updated Strategies for Monitoring and Assessment of Transboundary Rivers, Lakes and Groundwaters

UNITED NATIONS

Geneva, 2023

Contact information:
Water Convention secretariat
Email: water.convention@un.org
Website: www.unece.org/env/water

ECE/MP.WAT/70

UNITED NATIONS PUBLICATION
Sales No. E.23.II.E.1
ISBN 978-92-1-117316-1
eISBN: 978-92-1-002328-3

Photo credits:
Ministry of Water and Sanitation of Senegal: pages x, 24, 27, 35, 46, 57.
International Commission for the Protection of the Danube River: pages 41, 45.
Sirkku Tuominen, Finnish Environment Institute: page 49.
Zurab Jincharadze, EU Water Initiative Plus Project: pages 15, 20, 55.
Olexii Iaroshevich: page 16.
Vitaly Kolvenko: cover page top right.
UNECE: page 10.
Depositphotos: all other pictures.

Cover photos:
Left - Hydrogeological borehole
Top right - Hydropost on the Iagorlic River
Bottom right - Water monitoring station

FOREWORD

Finland and Senegal, as Co-Chairs of the Working Group on Monitoring and Assessment, are pleased to share the Updated Strategies for Monitoring and Assessment of Transboundary Rivers, Lakes and Groundwaters. This publication explains the key principles of and approaches to monitoring and assessment of transboundary waters that have been identified by governments and other stakeholders collaborating via the platform of the Convention on the Protection and Use of Transboundary Watercourses and International Lakes (Water Convention), which is hosted by the United Nations Economic Commission for Europe (UNECE).

The publication highlights policy options for decision-makers and provides ground rules for water managers involved in establishing and carrying out cooperation between riparian countries, as well as for representatives of joint bodies.

Water needs to be at the top of the global agenda in order to achieve the "ambitious and transformational vision" set out in the 2030 Agenda for Sustainable Development. Developing and sustaining joint monitoring and assessment programmes is indispensable for sustainable management of the 286 transboundary river and lake basins and 592 transboundary aquifer systems worldwide, and one of the core obligations of the Water Convention.

The opening of the Water Convention for accession by all United Nations Member States offers an important opportunity to advance the monitoring, assessment and management of transboundary waters at global scale. The present publication constitutes a tool to facilitate this work.

The Updated Strategies reflect global experience on monitoring, assessment and data sharing. The updates made in this edition give more explicit recognition to the link between water quality, groundwater and freshwater ecosystems and their ecology. The procedures for data management and data sharing are described in greater detail, and stronger emphasis is placed on the joint design of monitoring networks and joint sampling campaigns. This new edition also emphasizes the need for inclusiveness and participation of all stakeholders, as well as the need for capacity development at national and transboundary levels. New technologies and methodologies for monitoring that have emerged in the past decades are described and more attention is given to sustainability aspects of monitoring, including financing and sound legal frameworks.

We trust that this publication, issued on the eve of the United Nations 2023 Water Conference (New York, 22–24 March 2023), will provide valuable guidance to all actors involved in the management of transboundary waters and help them to improve understanding and implementation of the Water Convention. We invite countries and basins to use these Updated Strategies to strengthen transboundary cooperation and unlock the political, economic, social and environmental benefits that the sustainable management of our shared water resources can bring.

Anna-Stiina Heiskanen
Co-Chair of the Working Group
on Monitoring and Assessment
Finland

Niokhor Ndour
Co-Chair of the Working Group
on Monitoring and Assessment
Senegal

PREFACE

Joint monitoring and assessment of transboundary waters and information exchange are key obligations under the Convention on the Protection and Use of Transboundary Watercourses and International Lakes (Water Convention). For this reason, monitoring and assessment have been among the priority areas of work under the Water Convention already in the 1990s. The Working Group on Monitoring and Assessment was established in 2000, with workshops, training courses and other capacity building activities organised, pilot projects carried out and several guidance documents developed under its auspices. This work has led to the development of the Strategies for Monitoring and Assessment of Transboundary Rivers, Lakes and Groundwaters, adopted by the Meeting of the Parties to the Water Convention at its second session (Bonn, 20–22 November 2006).

At its ninth session (Geneva, 29 September – 1 October 2021) the Meeting of the Parties to the Water Convention entrusted the Working Group on Monitoring and Assessment to update the 2006 Strategies for Monitoring and Assessment of Transboundary Rivers, Lakes and Groundwaters to reflect the global experience and technological developments in this area.

The process of preparing the Updated Strategies for Monitoring and Assessment of Transboundary Rivers, Lakes and Groundwaters was coordinated by the lead Parties (Finland and Senegal) and the secretariat. The first draft was developed based on written consultations prior to and discussion at the Expert Meeting on Monitoring, Assessment and Data Exchange (1 April 2021), as well as inputs gathered during the Global Workshop on Exchange of Data and Information in Transboundary Basins (4–5 December 2019) and the Fifteenth meeting of the Working Group on Monitoring and Assessment (6 December 2019). The draft was then discussed at the Expert Meeting on Monitoring, Assessment and Data Exchange (13–14 April 2022) and revised based on feedback provided by experts during and after the meeting. The Fourth joint meeting of the Working Groups on Integrated Water Resources Management (IWRM) and on Monitoring and Assessment (28–30 June 2022, Tallinn, Estonia) reviewed the draft and entrusted the secretariat, in consultation with the lead Parties, with the task of integrating comments received. The secretariat, in cooperation with the lead Parties, then finalized the publication in November 2022.

The programme of work for 2022–2024 under the Water Convention, specifically activities devoted to "Supporting monitoring, assessment and information sharing in transboundary basins", aim to assist countries in initiating or further developing joint or coordinated monitoring and assessment of transboundary surface waters and groundwaters, covering both water quantity and quality aspects, and improving the exchange of information and data between riparian countries. The Updated Strategies for Monitoring and Assessment of Transboundary Rivers, Lakes and Groundwaters aim to contribute to the implementation of these activities.

ACKNOWLEDGEMENTS

This publication would not have been possible without the generous contributions of many governments, international organizations and individuals.

The Water Convention secretariat gratefully acknowledges the Co-Chairs of the Working Group on Monitoring and Assessment – Anna-Stiina Heiskanen (Director of the Freshwater Centre, Finnish Environment Institute, Finland) and Niokhor Ndour (Director of Water Resources Management and Planning at the Ministry of Water and Sanitation, Senegal) – and the former Co-Chair of the Working Group, Lea Kauppi (Director General, Finnish Environment Institute, Finland), for their guidance and inputs to the development of this publication.

It acknowledges the important work carried out by Jos Timmerman (Waterframes, advisor for the Dutch Ministry of Infrastructure and Water Management), who developed the text of this publication on the basis of the 2006 Strategies for Monitoring and Assessment of Transboundary Rivers, Lakes and Groundwaters, and subsequent comments received.

It also acknowledges the advice provided by the members of the Working Group on Monitoring and Assessment and the Expert Meetings on Monitoring, Assessment and Data Exchange (in 2021 and 2022), as well as work carried out on the development of this publication by numerous partners that contribute to implementation of the Convention's programme of work.

The secretariat wishes to thank all the participants to the online Expert Meeting on Monitoring, Assessment and Data Exchange (1 April 2021), which outlined the main directions for updating the 2006 Strategies for Monitoring and Assessment of Transboundary Rivers, Lakes and Groundwaters:

Mohammed Abdella Abdulkadir, Ethiopia; Mohamed Abdulrazzak, Saudi Arabia; Tommaso Abrate, World Meteorological Organization (WMO); Danael Aceves, Mexico; Asmaa Ait Bahassou, Morocco; Mohammed Al-Bakri, Iraq; Farida Alakbarova, Azerbaijan; Sarya Alameri, Iraq; Ahmed Alasadi, Iraq; Luay Aldaloo, Iraq; Alphonse Ahodègnon Alomasso, Benin; Ben Yaw Ampomah, Ghana; Ryad Awaja-Aouadja, State of Palestine; Ibrahima Ba, Organization for the Development of the Senegal River (OMVS); Macarena Bahamondes, Chile; Esi Biney, Ghana; Edoardo Borgomeo, World Bank; Corina Cosmina Boscornea, Romania; António Branco, Portugal; Maria Conagua Lagos, Mexico; Allie Davis, United States of America; Aboubakar Drabo, Burkina Faso; George Dzamukashvili, National Water Partnership of Georgia; Melchior Elsler, UNEP Global Environment Monitoring System for freshwater (GEMS/Water); Elisabeth Ericson, United States of America; Mona Fakih, Lebanon; Essam Fallatah, South Arabia; Amarildo Guri, Albania; Gilbert Gwati, Botswana; Seppo Hellsten, Finland; Tennielle Hendy, Belize; Muna Hindiyeh, German Jordanian University; Abdalraheem Huwaysh, Libya; Botir Ismoilov, Uzbekistan; Laith Jabbar, Iraq; Ameer Kadhim, Iraq; Merey Kasteyeva, Uzbekistan; Lea Kauppi, Finland; Ziad Khayat, Economic and Social Commission for Western Asia; Ulugbek Komilov, Uzbekistan; Peter Kovacs, Hungary; Viktor Lagutov, Central European University; Christina Leb, World Bank; Orozco Leon Samuel, CONAGUA; Nodirbek Madiboev, Uzbekistan; Jose Malanco; Pascal Mananga, Congo; Alberto Manganelli, Regional Centre for Groundwater Management (CeReGAS); Hayet Mansour, Tunisia; Elvira Marchidan, Romania; Carmen Marques Ruiz, European Commission; Zaida Martínez Casas, Mexico; Natalija Matic, independent researcher, Croatia; Zhanar Mautanova, International Water Assessment Center (IWAC); Maria Mihailescu, Romania; Olga Mirshina, Uzbekistan; Ylber Mirta, North Macedonia; Shafaq Mokwar, Sudan; Jasmine Moussa, Egypt; Eric Muala, Ghana; Patience Mukuyu, International Water Management Institute (IWMI); Maricela Munoz, Costa Ria; Ida Nagyné Sós, Hungary; Niokhor Ndour, Senegal; Aram Ngom, OMVS; Ainun Nishat, Bangladesh; Rosa Orient Quilis, Spain; Konstantina Papatsimpa, Greece; Tania Paredes, Mexico; Daniela Peña, Chile; Alina Petrenko, Ukraine; Nataliya Petrenko, Ukraine; Oleg Podolny, Hydrogeoecological Research and Design Company "KazHYDEC" (Ltd.), Kazakhstan; Sandra Poikane, EC Joint Research Centre; Ogopotsebatlokwa Pule, Botswana; Kamila Sabyrrakhim, Kazakhstan; Momin Saiyed, Afghanistan; Saniso Sakuringwa, Botswana; Maha Sall, OMVS; Silvia Saravia, Economic Commission for Latin America and the Caribbean; Kristina Schaufler, Environment Agency Austria; Andreas Scheidleder, Environment Agency Austria; Pedro Serra, Portugal; Mariia Shpanchyk, Ukraine; Gvantsa Sivsivadze, Georgia; Ulrika Stensdotter Blomberg, Sweden; Arnaud Sterckx, International Groundwater Resources Assessment Centre (IGRAC); Leyla Tagizadeh, Azerbaijan; Jos Timmerman, Netherlands; Callist Tindimugaya, Uganda; Ekaterina Veselova, Russian Federation; Steven Vinckier, Belgium; Sengphasouk Xayavong, Lao People's Democratic Republic; Mujtaba Zaka Ghulam, Afghanistan; Dinara Ziganshina, Scientific and Information Center of the Interstate Commission for Water Coordination of Central Asia (SIC ICWC); and Alexander Zinke, Austria.

The secretariat also wishes to thank all the participants to the Expert Meeting on Monitoring, Assessment and Data Exchange (13–14 April 2022), which reviewed the first draft of the Updated Strategies for Monitoring and Assessment of Transboundary Rivers, Lakes and Groundwaters:

Arailym Alimusina, IWAC; Abdoulaye Alio, Lake Chad Basin Commission; Alphonse Ahodègnon Alomasso, Benin; Juan Carlos Alurralde, Intergovernmental Coordinating Committee of the Countries of the La Plata Basin; Maria Apostolova, Amazon Cooperation Treaty Organization (ACTO); Radmila Bojkovska, North Macedonia; Corina Cosmina Boscornea, Romania; Martina Bussettini, Italy; Fernando Cisneros Arza, ACTO; Kilian Christ, UNEP; George Dzamukashvili, National Water Partnership of Georgia; Melchior Elsler, UNEP; Ikrom Ergashev, Uzbekistan; Samo Grošelj, International Sava River Basin Commission; Amarildo Guri, Albania; Anna-Stiina Heiskanen, Finland; Paul Haener, International Network of Basin Organizations (INBO); Mohamad Kayyal, UNEP/MAP; Peter Kovacs, Hungary; Ulugbek Komilov, Uzbekistan; Annukka Lipponen, Finland; Igor Liska, International Commission for the Protection of the Danube River (ICPDR); Elvira Marchidan, Romania; Natalija Matic, Croatia; Zhanar Mautanova, IWAC; Patience Mukuyu, IWMI; Laziz Nazariy, SIC ICWC; Oleg Podolny, Hydrogeoecological Research and Design Company "KazHYDEC" (Ltd.), Kazakhstan; Juan Carlos Muñoz, Peru; Vesa Puoskari, Finland; José Pérez, Dominican Republic; Anna Rautvuori, Permanent Mission of Finland in Geneva; Edwin Ruiz Ceballos, Dominican Republic; Claudia Ruz Vargas, IGRAC; Andreas Scheidleder, Environment Agency of Austria; Kristina Schaufler, Environment Agency of Austria; Gvantsa Sivsivadze, Georgia; Arnaud Sterckx, IGRAC; Hassane Tahirou, Lake Chad Basin Commission; and Leyla Tagizadeh, Azerbaijan.

In addition, the secretariat wishes to thank all participants to the Fourth joint meeting of the Working Groups on Integrated Water Resources Management (IWRM) and on Monitoring and Assessment (28–30 June 2022, Tallinn, Estonia) for their feedback to the draft of the Updated Strategies for Monitoring and Assessment of Transboundary Rivers, Lakes and Groundwaters.

Lastly, the Secretariat wishes to thank all the experts who provided written comments, both before and after the Expert Meeting on Monitoring, Assessment and Data Exchange (13–14 April 2022), and after the Fourth joint meeting of the Working Groups on Integrated Water Resources Management (IWRM) and on Monitoring and Assessment (28–30 June 2022, Tallinn, Estonia):

Paul Haener, INBO; Anna-Stiina Heiskanen, Director of the Freshwater Centre, Finnish Environment Institute and Co-Chair of the Working Group on Monitoring and Assessment; Turo Hjerppe, Senior Specialist, Finnish Ministry of Environment; Lea Kauppi, Secretary General of the Finnish Environment Institute and previous Co-Chair of the Working Group on Monitoring and Assessment; Jonathan Lautze, IWMI South Africa; Annukka Lipponen, Chief Specialist on Water Resources Management, Ministry of Agriculture and Forestry of Finland; Arshid Moh'd, Jordan; Niokhor Ndour, Director of Water Resources Management and Planning at the Ministry of Water and Sanitation of Senegal and Co-Chair of the Working Group on Monitoring and Assessment; Oleg Podolny, Hydrogeoecological Research and Design Company "KazHYDEC" (Ltd.), Kazakhstan; Kristina Schaufler, Environment Agency of Austria; Andreas Scheidleder, Environment Agency of Austria; Arnaud Sterckx, IGRAC; Eric Tardieu, INBO; Giacomo Teruggi, WMO; and Abrate Tommaso, WMO.

At the secretariat, Sara Datturi and Iulia Trombitcaia coordinated the preparation of the Updated Strategies for Monitoring and Assessment of Transboundary Rivers, Lakes and Groundwaters, supported by Sonja Koeppel. Annukka Lipponen coordinated the process of updating the publication as part of the secretariat up until June 2021. Komlan Sangbana, Chantal Demilecamps and Alexander Belokurov supported the Expert Meeting on Monitoring, Assessment and Data Exchange (13–14 April 2022). Minako Hirano and Cammile Marcelo provided administrative support throughout the process.

Finally, this publication would not have been possible without funding from Finland, Germany, the Netherlands, Sweden, Switzerland and the European Union.

While every effort has been made to name all contributors, the secretariat regrets if any individual or organization has been overlooked above.

CONTENTS

List of Figures

List of boxes

Monitoring, assessment and data collection activities in Senegal

1. INTRODUCTION

Information based on well-organized monitoring programmes is the key prerequisite for accurate assessments of the status of water resources and the magnitude of water problems. These assessments are essential for preparing policy actions at the local, national and transboundary levels to achieve associated goals and targets including those established by the 2030 Agenda for Sustainable Development. Moreover, integrated water resources management in transboundary basins shared by two or more countries calls for comparable information. There is a need for a common basis for decision-making, which requires comparable assessment methods and data management systems, as well as uniform reporting procedures. Data and information sharing, wide accessibility, fair and just sharing of open data and information, and joint monitoring and assessment also play an important role in building trust, thus facilitating cooperation and conflict avoidance. In this context, national governments play a crucial role in financing and developing shared databases, which are essential to the success of cross-border monitoring and assessment programmes.

This central role of data and information sharing in ensuring effective transboundary water cooperation has been recognized in the methodology for calculating Sustainable Development Goal (SDG) indicator 6.5.2, which measures the proportion of transboundary basin area with an operational arrangement for water cooperation. The existence of *at least annual exchanges of data and information* is one of four criteria for an arrangement to be considered operational.[1] However, the outcomes of the 2020 monitoring exercise under SDG indicator 6.5.2 and the Water Convention show that joint monitoring and exchange of data and information in transboundary river, lake and aquifer basins still represent a challenge for many countries.[2]

This publication explains key principles and approaches to monitoring and assessing transboundary waters and provides strategies to implement these processes. It also highlights areas of interest to policy and decision makers and provides ground rules for water managers involved in or responsible for establishing and carrying out cooperation between riparian countries, as well as for representatives of joint bodies.

The publication stresses the underlying legal, administrative and financial aspects of monitoring and assessment, and discusses the constraints on and opportunities for cooperation. It draws on experience gained from the implementation of pilot projects and lessons learned in monitoring, assessment and data sharing of transboundary rivers, lakes and groundwaters under the 1992 Convention on the Protection and Use of Transboundary Watercourses and International Lakes[3] (Water Convention). It then proposes step-by-step approaches for the development of monitoring, assessment and data sharing that take into account the available human and financial resources, including in countries facing economic challenges.

This publication builds on the 2006 Strategies for Monitoring and Assessment of Transboundary Rivers, Lakes and Groundwaters,[4] and the Guidelines on monitoring and assessment of transboundary rivers,[5] groundwaters[6] and lakes[7] developed under the Water Convention, as well as a review of other relevant international guidance[8] performed to assess the pertinence of the 2006 Strategies. The annexes provide overviews of specific aspects of groundwater, lake and river monitoring.

---

[1]   UN-Water, 2020. *Step-by-Step Monitoring Methodology for SDG Indicator 6.5.2*, available at https://www.unwater.org/publications/step-step-methodology-monitoring-transboundary-cooperation-652.

[2]   UNECE, UNESCO, 2021. *Progress on Transboundary Water Cooperation: Global Status of SDG Indicator 6.5.2 and Acceleration Needs*, available at www.unwater.org/publications/progress-on-transboundary-water-cooperation-652-2021-update

[3]   www.unece.org/env/water

[4]   UNECE, 2006. *Strategies for Monitoring and Assessment of Transboundary Rivers, Lakes and Groundwaters*, available at https://unece.org/DAM/env/water/publications/assessment/StrategiesM_A.pdf

[5]   UNECE Task Force on Monitoring and Assessment, 2000. *Guidelines on Monitoring and Assessment of Transboundary Rivers*, available at https://unece.org/DAM/env/water/publications/assessment/guidelines_rivers_2000_english.pdf

[6]   UNECE Task Force on Monitoring and Assessment, 2000. *Guidelines on Monitoring and Assessment of Transboundary Groundwaters*, available at https://unece.org/DAM/env/water/publications/assessment/guidelinesgroundwater.pdf

[7]   UNECE Working Group on Monitoring and Assessment, 2002. *Guidelines for the Monitoring and Assessment of Transboundary and International Lakes. Part A: Strategy document*, available at https://unece.org/DAM/env/water/publications/assessment/lakesstrategydoc.pdf and UNECE Working Group on Monitoring and Assessment, 2003. *Guidelines on Monitoring and Assessment of Transboundary and International Lakes. Part B: Technical guidelines, available at* https://unece.org/DAM/env/water/publications/assessment/lakestechnicaldoc.pdf

[8]   "Outlook for developing monitoring cooperation and exchange of data and information across borders: Background paper to the Global workshop on exchange of data and information and to the Fifteenth meeting of the Working Group on Monitoring and Assessment under the Water Convention (Geneva, 4–6 December 2019)", ECE/MP.WAT/WG.2/2019/INF.1, available at https://unece.org/environmental-policy/events/fifteenth-meeting-working-group-monitoring-and-assessment

Transboundary estuaries and other transitional waters (e.g. lagoons, deltas and coastal lakes) are not explicitly included in this publication, but should nevertheless be considered within the framework of the Water Convention, and the general principles and approaches described in this publication applied to these water bodies.[9] The specifics of estuaries, including tidal movements, hydrological regime and salinity, require a targeted approach for monitoring practice and are covered in Annex 4.

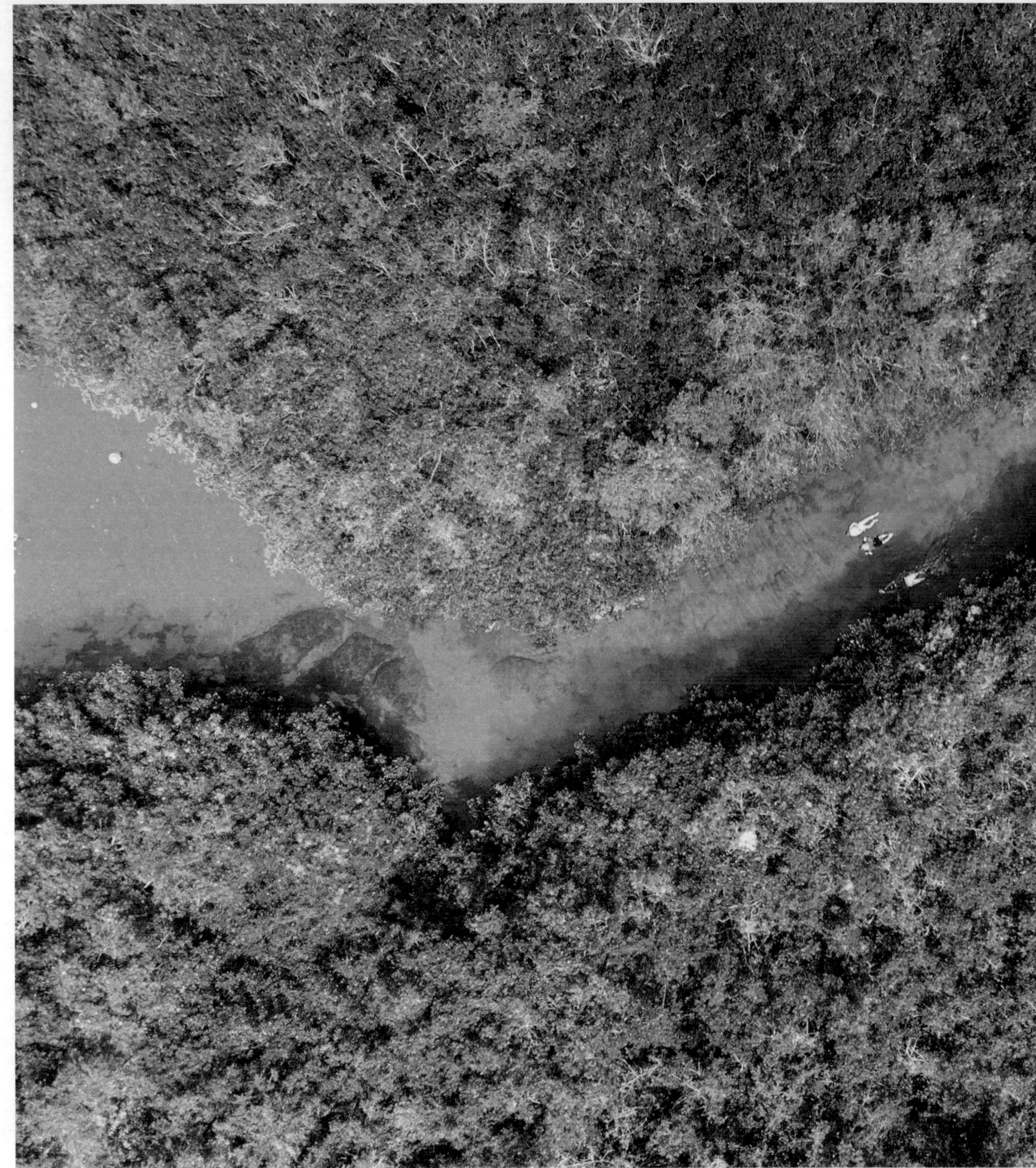

---

[9]   See also Working Group on Monitoring and Assessment, 2022. *Working paper on guidelines on monitoring and assessment of transboundary estuaries*, available at https://unece.org/DAM/env/water/meetings/wgma/doc/wgma-2002-7.pdf and Common Implementation Strategy for the Water Framework Directive (2000/60/EC), Guidance document No. 7. *Monitoring under the Water Framework Directive*, available at https://circabc.europa.eu/sd/a/63f7715f-0f45-4955-b7cb-58ca305e42a8/Guidance%20No%207%20-%20Monitoring%20(WG%202.7).pdf

Casa Cenote, Mexico

Old countryside water well with pulley and bucket in Thailand

2. BASIC PRINCIPLES AND APPROACHES

Summary

This chapter describes the basic principles and approaches of monitoring and assessment in a transboundary context, and how they support decision-making. It explains that a monitoring and assessment approach should seek to identify the sources of pressure on the water system and evaluate their impacts and severity. It should also determine the overall status of the water system, focusing on utilization, with a view to implementing relevant measures. The monitoring system should be inclusive and gender responsive and should adopt a basin approach, taking into account the different purposes for which data will be used. The chapter also presents the benefits of joint monitoring.

2.1 Monitoring and assessment

The ultimate goal of monitoring[10] is to provide the information necessary for planning, decision-making and operational water management at the local, national and transboundary levels. Monitoring programmes are also fundamental to the protection of human health and the environment in general. Assessment is a crucial part of monitoring[11] because it translates the acquired data into information about the current state of a water body. It provides the basis for describing changes and trends which can be linked to pressures and impacts and then related to environmental targets or objectives (Figure 1). Assessment also encompasses boundary conditions and the broader social and environmental context determining the state of the environment.

Identifying, documenting and prioritizing the various uses and functions of a surface water or groundwater basin, and the associated water management issues, is central to the development of a useful monitoring and assessment programme. The Driving forces–Pressures–State–Impact–Responses (DPSIR) framework can help clarify the connections between different water management issues (Figure 1).

Figure 1: Driving forces–Pressures–State–Impact–Responses (DPSIR) framework

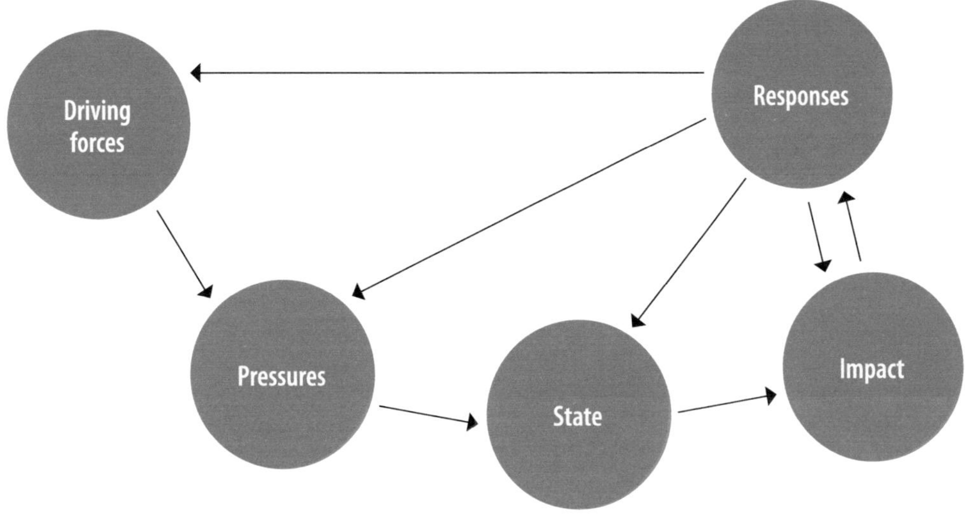

Source: EEA, 1998. *Europe's Environment: The Second Assessment.* Elsevier Science Ltd., Oxford, UK.

---

[10] Monitoring is a process of systematically scrutinizing and checking with a view to collecting data (www.wef.org/resources/for-the-public/public-information/glossary).

[11] Assessment involves the determination of sources, extent, dependability and quality of water resources, and of the human activities that affect those resources, for their utilization and control (https://community.wmo.int/activity-areas/water-resources-assessment).

The DPSIR framework assumes that social, economic and environmental systems are interrelated and shows how drivers of environmental change create pressures on the environment, which in turn affect the state of the environment resulting in impacts on ecosystems, economies and communities. Negative impacts will eventually lead to responses by society, including the development of policies for basin protection. If a policy has the intended effect, its implementation will influence the drivers, pressures, state and impacts.

Analysis of information needs is the most critical step in developing a successful, tailor-made and cost-effective monitoring programme. In general, information will be required on each of the elements in the DPSIR framework.

Monitoring is usually understood as a process of repetitive measurement and observation conducted for various defined purposes and focused on one or more elements of the environment. To ensure comparability over time, the monitoring processes are undertaken at the same locations at regular time intervals, using comparable methodologies for environmental sensing and data collection.

Monitoring enables assessment of the current state of water quantity and quality including their variability in space and time. Such assessments often appraise hydrological, morphological, physicochemical, chemical, biological and/or microbiological conditions in relation to reference conditions, human health effects and/or existing or planned uses of water. Such reference conditions comprise the natural variability of geophysical and geochemical processes that may influence concentrations of specific variables.

A specific purpose of monitoring is to support decision-making and operational water management in critical situations. For example, in critical hydrological situations such as floods, ice drifts and droughts, the availability of timely and reliable hydro-meteorological data is essential and often requires telemetric systems to ensure the continuous transmission of such information. Reliable data are also needed in the case of pollution events, which may require early warning systems to signal when critical pollution levels are exceeded, or toxic effects occur. In such cases, models can often support decision-making.

Since monitoring and assessment is essential for preparing proper policy actions, the monitoring system needs to be gender responsive and inclusive, and the data and information accessible. Identifying the factors that contribute to the inclusion or exclusion of women and men belonging to different social, cultural or ethnic groups, such as Indigenous people, and the ways in which they interact with water resources for different uses, could improve the provision, management and conservation of the world's water resources for the benefit of all. Representation of diverse groups of stakeholders at all stages of the monitoring cycle is a basic starting point to this end.[12] Moreover, gender-disaggregated statistics are essential to elucidate the circumstances, life situation and needs of men and women.[13]

For transboundary waters, information is often gathered from national monitoring systems – which are established and operated according to national laws and regulations and international agreements – rather than monitoring systems specifically established and operated by joint bodies. Thus, national legislation, as well as obligations under international agreements and other commitments, should be carefully examined in preparation for establishing, upgrading and running these systems.

2.2 Basin approach

The basin forms a natural unit for integrated water resources management in which rivers, lakes and groundwaters interact with other ecosystems. A basin refers to either the area of land from which all surface runoff flows through a sequence of streams, rivers, groundwater bodies and possibly lakes into the sea at a single river mouth, estuary, lagoon or delta (Figure 2), or the area of land from which all surface runoff ends up in another final recipient of water, such as a lake or a desert. The whole basin should therefore be considered when developing a monitoring system.

As basins usually stretch over different administrative and geographical units and state borders, cooperation between competent actors is essential. Among these actors are environmental and water agencies, hydro-meteorological services, geological surveys, public health institutions and water laboratories. They also include research institutes and universities engaged in methodological work on monitoring, modelling, forecasting and assessments. In addition, the knowledge of local and Indigenous people groups should be incorporated into monitoring systems.[14] Such cooperative arrangements and institutional frameworks greatly influence the efficiency of monitoring and

---

[12] See, for example, the UNESCO, 2019. *Gender-Responsive Indicators for Water Assessment, Monitoring and Reporting*, available at https://unesdoc. unesco.org/ark:/48223/pf0000367971.locale=en

[13] See, for example, www.includegender.org/toolbox/map-and-analyse/gender-statistics

[14] UNESCO, 2021. *Assessing and Certifying Indigenous Tracking Expertise and Skills*, available at https://en.unesco.org/sites/default/files/links_indigenous_tracking_20210913.pdf

Figure 2: Key elements of the hydrological cycle of a basin

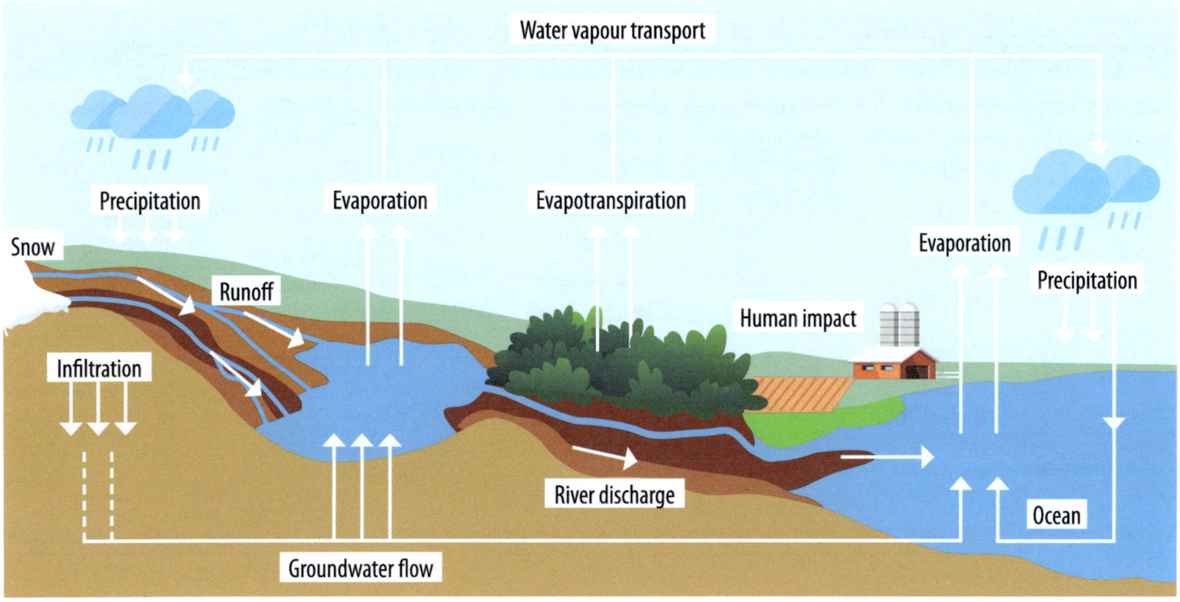

Source: UNECE, *Strategies for Monitoring and Assessment of Transboundary Rivers, Lakes and Groundwaters* (New York and Geneva, United Nations, 2006).

assessment. Moreover, concerted action plans, as required under the Water Convention, and water resources management plans are an important basis for specifying the information needs for monitoring and assessment.

The level of detail that monitoring and assessment can provide depends on the density of the network, the frequency of measurements/observations, the size of the basin and/or the issues under investigation. For example, when a measuring station at the outlet of a (sub-)basin reports water-quality changes, a more detailed monitoring network often will be necessary to determine the sources, causal agents and pathways of pollutants. The interactions between surface waters and groundwaters may also be different in the upper and lower parts of the basin. In such cases, information is needed for smaller sub-basins. Monitoring networks, frequency of measurements and variables, and assessment methodologies should all be adapted to meet these conditions. A conceptual model of the basin should also be developed in order to reveal the interactions between surface water and groundwater and between water quantity and quality.

2.3 Different purposes

Information based on well-organized monitoring programmes that account for the complexity of issues (Figure 3) is a prerequisite for accurate assessments of the status of water resources and the magnitude of water problems. Furthermore, such assessments are essential for preparing proper policy actions at the local, national and transboundary levels. At the transboundary level there is a need for a common basis for decision-making, which requires harmonized and comparable data and information. Indeed, water resources management in transboundary basins requires sharing data and information that meets the expectations of stakeholders for various activities. Figure 3 outlines some of the main domains requiring access to water-related data.

The regular exchange of data and information is also fundamental to establishing good cooperation between countries. This is particularly important for routine water resource operational management such as water sharing for irrigation and natural living resources (e.g. migratory fish and other assets based on biodiversity), as well as for medium or long-term basin planning, with monitoring of the programme of measures and investments. Unfortunately, in many cases, data collection processes and data sharing are limited, and when data are available, existing datasets are usually fragmented, incomplete, dispersed and heterogeneous. When developing and maintaining monitoring and data-sharing systems, it is essential that the whole information system is supported by an appropriate institutional framework. This includes clearly delineating the responsibilities of each actor and ensuring that sustainable funding and resources are accessible. Assigning and sharing responsibilities is of particular importance in a transboundary setting.

Figure 3: Different purposes for water data

Water data and information management are essential for:						
Sectorial water management	**Integrated Water Resource Management**	**Climate change adaptation and disaster risk reduction**	**Decision making**	**Reporting**	**Specific decision taking**	**Other water activities**
• Ecosystem/ environment	• Local level	• Floods	• Development of policies and legislation	• Global	• Operational management	• Regulatory aspects
• Drinking water supply	• Basin level	• Shortages	• Assessment of policy impacts	• Regional	• Territory management	• Partners
• Agriculture	• National level	• Drought	• Surveillance of policy implementation	• National statistics	• Emergency situations	• Public information
• Energy	• Transboundary basins			• Specific Conventions		
• Health	• Regional level					
• Transportation						

Source: Modified from INBO, 2018. *The Handbook on Water Information Systems Administration: Processing and Exploitation of Water-related Data.*

2.4 Benefits of joint monitoring

Joint monitoring includes substantial benefits for countries. During the second reporting exercise under the Water Convention in 2020, countries were asked to report on the main achievements they experienced in relation to joint monitoring. A range of benefits and achievements was mentioned including:

- mutual support in establishing a monitoring system, developing a joint approach to the future proposal of measures, optimization of activities, joint capacity building, implementing a shared database and drafting joint studies;

- agreement on and approvement of monitoring parameters and methods, and harmonization of results from chemical, ecological and biological analysis of water from agreed monitoring stations;

- improved basin-wide, transparent, harmonized, "neutral" and reliable information, and data on the state of the environment leading to greater technical and scientific understanding of the entire basin as the basis for better management of water bodies;

- improved forecasting, impact assessment and dissemination of results for better decision-making;

- the development of regular reports such as impact studies and state of the basin reports;

- improved early warning through the availability of continuous monitoring results to detect contaminations in time for intervention, as well as for flood forecasting and disaster risk management, including successful coordination and cooperation during flooding events;

- improved understanding of the distribution of a basin's water resources and water balance, enabling the setting of environmental flows, better control and operational rules for the basin and sub-basins, and efficient water supply to parties involved;

- shared concepts of pressures and impacts providing a common ground for cooperation, offering a platform for dispute settlement and improved trust and confidence among riparian states, their institutions, citizens and Indigenous people, as well as enhanced cooperation.

Mekong River delta, Can Tho, Viet Nam

UNECE
Session 4: Legal Basis and Institutional Framework for transboundary data and information exchange
Progressive evolution of the legal basis for data and information exchange in the Sava River Basin: protocols, data policy, GIS and information systems
Mirza Sarač
Advisor, ISRBC Secretariat

3. LEGISLATION AND COMMITMENTS

Summary

Multilateral environmental agreements including various Conventions and Protocols, as well as bilateral and multilateral transboundary water agreements, contain obligations for countries to monitor and assess waters, and to report, as appropriate, to a specific body, such as an international commission, secretariat or organization. Ideally, these obligations should be introduced into national legislation to steer activities in national competent bodies.

A legal framework at national and basin level is imperative in order to set up and maintain a transboundary monitoring and assessment system. In addition, national legislation should set out obligations and responsibilities for relevant agencies, such as hydro-meteorological services, environmental and health agencies, geological surveys, and operators of water regulation structures and industrial installations, to monitor and assess the various components of the environment and report on the results.

This chapter covers several global and regional instruments dealing with environmental data and information, among others.

3.1 Global instruments

3.1.1 Water Convention

The main goal of the 1992 Convention on the Protection and Use of Transboundary Watercourses and International Lakes (Water Convention)[15] is to prevent, control and reduce any transboundary impacts that include significant adverse effects on human health and safety, flora, fauna, soil, air, water, climate, landscape, and historical monuments or other physical structures. The Convention is one of the most essential legal instruments for the monitoring and assessment of transboundary waters.

In regard to defining and specifying information needs, establishing monitoring systems and assessing the status of waters, the Convention requires that emission limits be set for discharges from point sources on the basis of the best available technology (BAT). It also requires authorizations for wastewater discharges and the application of at least biological or equivalent processes to treat municipal wastewater.

In addition, the Convention calls for best environmental practices (BEP) to reduce the input of nutrients and hazardous substances from agriculture and other diffuse sources. In addition, Parties must define water-quality objectives for the purpose of preventing, controlling and reducing transboundary impacts.

The Convention further requires the establishment and implementation of joint programmes to monitor the condition of transboundary waters and transboundary impacts, as well as joint or coordinated assessments made at regular intervals and exchange of data and information.

Obligations relating to the monitoring and assessment of specific basins that stem from bilateral or multilateral agreements should align with the requirements of the Water Convention. In particular, joint bodies – any bilateral or multilateral commission or other appropriate institutional arrangements for cooperation between Riparian Parties – play a specific role in monitoring and assessment.

3.1.2 Watercourses Convention

The 1997 Convention on the Law of the Non-navigational Uses of International Watercourses[16] (Watercourses Convention) aims to ensure the sustainable utilization of international watercourses in an equitable and reasonable manner. In general, Parties to the Watercourses Convention have the obligation to cooperate and not to cause significant harm. Pursuant to this, Parties must exchange data and information on the condition of the watercourse and related planned measures on a regular basis, as well as on the request of another riparian Party.

The Watercourses Convention and the Water Convention are fully compatible. There is no contradiction between the two instruments, and a country can be a Party to both Conventions.

---

15   https://unece.org/environment-policy/water

16   https://treaties.un.org/doc/Treaties/1998/09/19980925%2006-30%20PM/Ch_XXVII_12p.pdf

3.1.3 Other global instruments

Legal obligations regarding the monitoring and assessment of transboundary waters also arise from other international legal instruments, such as the Convention on Wetlands of International Importance especially as Waterfowl Habitat (Ramsar Convention),[17] the Convention on Biological Diversity[18] and the Convention to Combat Desertification.[19]

3.2 Regional instruments

3.2.1 Protocol on Water and Health

Under the 1999 Protocol on Water and Health[20] to the 1992 Water Convention, effective systems should be established to monitor and assess situations likely to result in outbreaks or incidents of water-related disease, and to respond to or prevent their occurrence. This may include inventories of pollution sources, surveys on high-risk areas for microbiological contamination and toxic substances, and reporting on infectious and other water-related diseases. The Parties must also develop integrated systems to handle information about long-term trends in water and health; current concerns, past problems and successful solutions; and the provision of such information to the authorities. Moreover, comprehensive national and/or local early warning systems must be established, improved or maintained.

3.2.2 Industrial Accidents Convention

The 1992 Convention on the Transboundary Effects of Industrial Accidents[21] is designed to protect human beings and the environment against industrial accidents by preventing them to the extent possible, reducing their frequency and severity, and mitigating their effects. Through prevention, preparedness and response to industrial accidents, the Convention also supports disaster risk reduction.

To respond effectively and in a coordinated way to an industrial accident, Parties must be informed as soon as possible, since time is of essence. The Convention consequently calls on Parties to set up special notification systems. The UNECE Industrial Accidents Notification System, developed under the Convention, comprises a network of points of contact and enables prompt notification of all potentially affected countries in the event of a major accident with transboundary effects, including in the cases of accidental pollution on water bodies.

3.2.3 Aarhus Convention

The 1998 Convention on Access to Information, Public Participation in Decision-making and Access to Justice in Environmental Matters[22] (Aarhus Convention) stipulates, inter alia, that any environmental information held by a public authority must usually be provided when requested by a member of the public. The scope of information is quite broad, and includes information on water and human health and safety. Public authorities may impose a charge for supplying the requested information provided the charge does not exceed a "reasonable" amount. There is also an obligation to progressively make environmental information publicly available via electronic databases. The Convention specifies certain categories of information (e.g. state of the environment reports) that should be made available in this form.

The 2003 Protocol on Pollutant Release and Transfer Registers (PRTRs)[23] to the Aarhus Convention requires Parties to establish and maintain a publicly accessible national PRTR with information on releases of pollutants in the air, water and land. The information contained in the PRTR is to be supplied through mandatory periodic reporting by the owners or operators of polluting facilities. The Protocol requires that PRTRs also progressively contain information on pollution from diffuse sources, such as from agriculture to water.

---

[17]   www.ramsar.org

[18]   www.cbd.int/convention

[19]   www.unccd.int

[20]   https://unece.org/environment-policy/water

[21]   https://unece.org/environment-policy/industrial-accidents

[22]   www.unece.org/env/pp/welcome.html

[23]   https://unece.org/env/pp/protocol-on-prtrs-introduction

3.2.4 Escazú Agreement

The 2018 Regional Agreement on Access to Information, Public Participation and Justice in Environmental Matters in Latin America and the Caribbean,[24] better known as the Escazú Agreement, is an international treaty signed by 24 Latin American and Caribbean nations concerning the rights of access to information about the environment, public participation in environmental decision-making, environmental justice, and a healthy and sustainable environment for current and future generations.

The Escazú Agreement is the first international treaty in Latin America and the Caribbean related to the environment. It aims to provide full public access to environmental information and decision-making, and legal protection and recourse concerning environmental matters.

3.3 Other international commitments

Commitments on monitoring and data sharing are present in many other regional, subregional, basin-wide and bilateral agreements on transboundary cooperation. This section does not provide details of individual basin or bilateral agreements but does describe a few international commitments that cover water management at a regional or subregional scale.

3.3.1 Amazon Cooperation Treaty Organization (ACTO)

The Amazon Cooperation Treaty Organization (ACTO), an intergovernmental organization formed by the eight Amazonian countries: Bolivia, Brazil, Colombia, Ecuador, Guyana, Peru, Suriname and Venezuela, has developed a strategic agenda for the conservation and sustainable use of renewable natural resources and sustainable development. The "knowledge management and information sharing" section of this Amazon Strategic Cooperation Agenda[25] emphasizes the exchange of information, knowledge and technology on all thematic areas, including water resources, under the principles of solidarity, reciprocity, respect, harmony, complementarity, transparency, equilibrium and equitable conditions. Information flows between the institutions and intergovernmental authorities of Member Countries are fostered by the Amazon Regional Observatory.[26]

3.3.2 EU legislation

The legislation of the European Union is a major tool for defining how surface waters and groundwaters should be used, protected and restored in the EU region. The key directive concerning monitoring is Directive 2000/60/EC of the European Parliament and of the Council of 23 October 2000, which establishes a framework for European Community (EC) action in the field of water policy, hereinafter referred to as the Water Framework Directive (WFD).[27] The WFD provides the framework for the protection of surface waters, transitional waters, coastal waters and groundwaters in the EU area. The main aims of the WFD are to prevent further deterioration of and protect and enhance the status of aquatic ecosystems, to promote their sustainable water use, and to mitigate the effects of floods and droughts. The overriding environmental objective is to prevent deterioration of the status of all waters and to achieve good ecological and chemical status for waters by 2027 at the latest.

Within river basins where use of water may have transboundary effects, the requirements for achievement of environmental objectives established under the WFD, and in particular all programmes of measures, should be coordinated for the whole basin. For river basins extending beyond the boundaries of the European Community, member states should endeavour to ensure appropriate coordination with the relevant non-member states. The WFD is meant to contribute to the implementation of EC obligations under international conventions on water protection and management, notably the Water Convention. Through its unifying role, and in particular through joint River Basin Management Plans (RBMPs), the WFD has supported the harmonization and intercalibration of approaches, indicators and standards.

Under the WFD, a water monitoring system must be established to provide a coherent and comprehensive overview of the ecological and chemical status of each basin. To address the challenges in a cooperative and coordinated way, member states, Norway and the European Commission agreed upon a Common Implementation Strategy. A number of guidance documents[28] covering, among other issues, monitoring and public participation were prepared to support implementation of the WFD.

---

[24] https://repositorio.cepal.org/bitstream/handle/11362/43583/1/S1800428_en.pdf

[25] http://otca.org/en/wp-content/uploads/2021/01/Strategic-Agenda-of-Amazon-Cooperation.pdf

[26] https://oraotca.org/en

[27] https://eur-lex.europa.eu/legal-content/EN/TXT/?uri=CELEX%3A32000L0060&qid=1643807539361

[28] https://ec.europa.eu/environment/water/water-framework/facts_figures/guidance_docs_en.htm

Underlying the WFD is the Directive on Environmental Quality Standards (Directive 2008/105/EC) (EQSD),[29] also known as the Priority Substances Directive, which establishes environmental quality standards (EQS) for substances in surface waters. The list is regularly updated.

The Drinking Water Directive (Council Directive 98/83/EC of 3 November 1998 on the quality of water intended for human consumption) is intended to protect human health from the adverse effects of any contamination of water intended for human consumption by ensuring that it is wholesome and clean. In 2020, the Directive[30] was revised to reinforce water quality standards through the inclusion of emerging pollutants and the introduction of a preventive approach that favours actions to reduce pollution at its source.

The Groundwater Directive (2006/118/EC)[31] establishes a regime which sets groundwater quality standards and introduces measures to prevent or limit inputs of pollutants into groundwater. The directive also establishes quality criteria that take into account local characteristics and allow for further improvements to be made based on monitoring data and new scientific knowledge.

Directive 2007/60/EC on the assessment and management of flood risks entered into force on 26 November 2007,[32] and aims to reduce and manage risks that floods pose to human health, the environment, cultural heritage and economic activity. The Floods Directive requires Member States to assess all water courses and coastlines at risk from flooding, map the flood extent and assets and humans at risk in these areas, and take adequate and coordinated measures to reduce such risks. The Directive also reinforces public rights to access this information and have a voice in the planning process.

Directive (EU) 2019/1024 on open data and the re-use of public sector information[33] stipulates minimum requirements for EU member states regarding making public sector information available for re-use, and provides a common legislative framework for this area.

3.3.3 African Ministers' Council on Water (AMCOW)

The African Ministers' Council on Water (AMCOW) has introduced a harmonized process for monitoring and reporting on water and sanitation targets across several international agreements. A web-based reporting system, the Pan-African Water and Sanitation Sector Monitoring and Reporting System (WASSMO),[34] supports this process.

The platform aims to facilitate a comprehensive and harmonized approach to building monitoring capacity in the region. It provides for reporting on seven themes: water infrastructure for growth; managing and protecting water resources; water supply, sanitation, hygiene and wastewater; climate change and disaster risk reduction; governance and institutions; financing; and information management and capacity development.

3.3.4 Southern African Development Community (SADC)

The SADC Revised Protocol on Shared Watercourses[35] was developed in 2000 with the overall objective of fostering closer cooperation for judicious, sustainable and coordinated management, protection and utilization of shared watercourses, and to advance the SADC agenda of regional integration and poverty alleviation. The Protocol establishes a framework for the utilization of watercourses shared by two or more SADC member states, and provides for the establishment of shared Watercourse Institutions which will furnish all information necessary to assess progress on implementation of the Protocol.

Guidelines on Strengthening River Basin Organizations[36] developed under the SADC Revised Protocol emphasize the importance of implementing information exchange between the respective governments.

---

[29] https://eur-lex.europa.eu/legal-content/EN/ALL/?uri=CELEX:32013L0039

[30] https://eur-lex.europa.eu/eli/dir/2020/2184/oj

[31] https://eur-lex.europa.eu/legal-content/EN/TXT/?uri=CELEX:02006L0118-20140711

[32] https://eur-lex.europa.eu/legal-content/EN/TXT/?uri=CELEX:32007L0060

[33] https://eur-lex.europa.eu/legal-content/EN/TXT/PDF/?uri=CELEX:32019L1024&from=EN

[34] https://amcow-online.org/initiatives/water-and-sanitation-sector-monitoring-and-reporting-system-wassmo

[35] www.sadc.int/files/3413/6698/6218/Revised_Protocol_on_Shared_Watercourses_-_2000_-_English.pdf

[36] www.sadc.int/files/4513/5333/8265/SADC_guideline_establishment.pdf

3.3.5 Regional seas conventions

Several regional seas conventions[37] have protocols that concern the assessment and prevention of pollution from land-based sources, especially as riverine transport and land-based activities can be a major source of pollutants to estuaries, other transitional waters and adjacent marine waters. Obligations and guidelines under these protocols relevant to transboundary transitional waters include the monitoring and assessment of pressures, such as the riverine loading of nutrients and harmful substances and the environmental state of transitional and coastal waters.

Pilot monitoring exercise Armenia-Georgia, September 2020

---

[37] This refers to, among others, the Convention for the Protection of the Mediterranean Sea Against Pollution (Barcelona Convention), the Convention for the Protection and Development of the Marine Environment in the Wider Caribbean Region (Cartagena Convention), the Convention for Cooperation in the Protection, Management and Development of the Marine and Coastal Environment of the Atlantic Coast of the West and Central Africa Region (Abidjan Convention), the Nairobi Convention for the Protection, Management and Development of the Marine and Coastal Environment of the Western Indian Ocean (Nairobi Convention), the Framework Convention for the Protection of the Marine Environment of the Caspian Sea (Tehran Convention), the Convention on the Protection of the Marine Environment of the Baltic Sea Area establishing the Baltic Marine Environment Protection Commission (Helsinki Commission – HELCOM) and the Convention for the Protection of the Marine Environment of the North-East Atlantic (the "OSPAR Convention").

Moldovan and Ukrainian experts fill in the hydromorphological field study protocol in the framework of the Working Group on Monitoring and Data Exchange under the Dniester Commission

4. ESTABLISHING THE INSTITUTIONAL FRAMEWORK

Summary

This chapter presents institutional arrangements at the national and transboundary levels that are pre-conditions to ensuring cooperation among various governmental entities, the private sector and others. The chapter also describes institutional arrangements related to quality control procedures, and frameworks for exchanging and accessing information.

4.1 Institutional arrangements at the national level

Suitable institutional arrangements at the national and local level are a precondition for monitoring and assessment of transboundary waters, as they ensure cooperation among various governmental entities, the private sector and others. When making these arrangements, it is important to note that responsibility for groundwater monitoring and assessment with regard to water quality and quantity may lie with geological survey organizations rather than environmental or water agencies. Conversely, environmental agencies can often supply data on the ecological and biophysical parameters of waterbodies including ecological status, biodiversity, hydro-morphology, land degradation, waste and so on. Therefore, ample attention should be paid to the capacity development of all individuals involved.

Coordination of monitoring and assessment at the national level is a prerequisite to ensuring effective and efficient water management. This collaboration includes the various organizations involved in water management, including basin agencies. Moreover, cooperation among water, environmental and health authorities is vital to ensure the collection and use of data related to human health and safety.

Hydro-meteorological services play an essential role in providing water quantity data and early warning information for extreme hydrological events. Organizations which operate response systems for emergencies involving water regulation structures and industrial plants are important partners in providing data to mitigate the adverse impacts of failures or other accidents at installations on transboundary waters. Industrial enterprises that monitor their own water abstractions and wastewater discharges also provide data for compliance purposes. The assessment of watercourses also requires socio-economic data, including population and economic statistics, which are collected by statistical offices. In many instances, it is necessary to seek expertise from research institutions, universities or the private sector.

4.2 Institutional arrangements at the transboundary level

As noted above, functioning institutions and suitable institutional arrangements for monitoring and assessment at the national and local levels are a prerequisite for international cooperation. This is particularly the case in connection with the work of joint bodies, which includes the implementation of monitoring- and-assessment-related tasks. The joint body should function as a forum for the exchange of information and data, including on planned measures and activities, and for the harmonization of monitoring approaches.[38] Particular effort therefore should be made to build and strengthen their capacity.

Riparian countries may decide to establish a specific working group under the joint body, in which experts from different disciplines meet regularly to agree upon the implementation of monitoring and assessment activities, including technical, financial and organizational aspects.

Basic requirements for joint monitoring and assessment which might be set out in an arrangement, annex or protocol include: coordinated or harmonized data gathering and processing methods, databases, digitalization of data, provision of access to information online, compatibility of laboratories taking part in the monitoring, joint research and studies, exchange of knowledge and use of models, monitoring arrangements (regulations), and coordinated or harmonized monitoring and assessment programmes.[39] In the absence of a joint body, riparian countries can decide to establish an arrangement specifically for monitoring and assessment.

---

[38] UNECE, 2018. *Principles for Effective Joint Bodies for Transboundary Water Cooperation under the Convention on the Protection and Use of Transboundary Watercourses and International Lakes*, available at https://unece.org/DAM/env/water/publications/WAT_Joint_Bodies/ECE_MP.WAT_50_Joint_bodies_2018_ENG.pdf

[39] UNECE, 2021. *Practical Guide for the Development of Agreements or Other Arrangements for Transboundary Water Cooperation*, available at https://unece.org/sites/default/files/2021-11/ece_mp.wat_68_eng.pdf

While monitoring systems usually operate at a national level, some function at a transboundary level through a basin or sub-basin arrangement. More information about data and information exchange and the role of joint bodies can be found in the reports[40] on SDG indicator 6.5.2 and on progress in implementation of the Water Convention.

Joint monitoring exercises are a good means to improve harmonization of information across riparian countries. Such exercises can be conducted at intervals supplemental to regular national monitoring.[41]

In transboundary monitoring, it is advisable to seek the involvement of border guards to facilitate joint sampling near the borderline, the transport of samples across the border and timely delivery of samples to laboratories.

4.3 Institutional arrangements related to quality control procedures

Quality control procedures are essential to ensure the reliability of information obtained by monitoring. The quality system should incorporate all elements of the monitoring and assessment cycle, starting with documenting procedures for the specification of information needs and the development of an information strategy. Standards, established under the auspices of the International Organization for Standardization (ISO), the European Committee for Standardization (CEN), and other organizations for sample collection, transport and storage, and laboratory analysis, form the basis of the quality system. The World Meteorological Organization (WMO), as a standard-setting organization, has also developed a series of hydro-meteorological guidelines and regulations. Protocols for data validation, storage and sharing, as well as data analysis and reporting, should be established and documented.[42] Where appropriate, riparian countries should assign responsibilities related to quality systems to the joint bodies or via the joint arrangement. In this context, transboundary cooperation at the local level should be encouraged and promoted, including direct contacts between the laboratories and institutions involved.

As many decision makers are unaware of quality control procedures, it is important to use a step-by-step approach when strengthening quality assurance. The process will start with simple internal quality control measures, then proceed to overall accreditation and finally to international standards.[43] Quality management (QM) – incorporating quality assurance and quality control – has four major benefits:

- It enables better management of the process and a more effective organization.
- It leads to employee satisfaction and commitment to the organization.
- It improves the quality of products and services.
- It improves customer satisfaction and the image of the hydrological services.

Implementation of quality management systems will assist hydrological services in the provision of good management practices and ultimately will enhance confidence in the quality of their data, products and services. The quality management process forms part of the Monitoring and Assessment Cycle (Figure 4) and includes the following elements:

- definition of goals (monitoring, management, environmental, etc.);
- information requirements (including acceptable uncertainty);
- a holistic value chain approach (QM is embedded across the whole system);
- the selection of variables to be monitored;
- processes (including data rescue and validation);
- data handling and management; and
- institutional arrangements in support of QM implementation.

---

[40] UNECE, UNESCO, 2021. *Progress on Transboundary Water Cooperation: Global Status of SDG Indicator 6.5.2 and Acceleration Needs*, available at https://unece.org/environment-policy/publications/progress-transboundary-water-cooperation-global-status-sdg and UNECE, 2021. *Progress on Transboundary Water Cooperation under the Water Convention: Second Report on Implementation of the Convention on the Protection and Use of Transboundary Watercourses and International Lakes, 2017–2020*, available at https://unece.org/info/publications/pub/360105

[41] A good example of a joint monitoring exercise is the Joint Danube Survey (www.danubesurvey.org/jds4)

[42] UNECE Task Force on Monitoring and Assessment, 1996. *Quality Assurance*, available at https://unece.org/DAM/env/water/publications/documents/quality_assurance.pdf

[43] See, for example, ISO/IEC/EN 17025 covering general requirements for the competence of calibration and testing laboratories www.fasor.com/iso25

4.4 Frameworks for exchanging and accessing information

According to provisions of the Aarhus Convention and the Escazú Agreement, environmental information must be available to the public. According to the provisions of the Water Convention, riparian countries should exchange relevant information on surface water and groundwater quality and quantity. Finally, according to SDG indicator 6.5.2, an operational arrangement between riparian countries requires a regular (at least once per year) exchange of data and information (criterion 4).

Arrangements for the exchange of information among riparian countries should be governed by rules jointly agreed by these countries. The arrangements should specify the format and frequency of exchange, with readily available information exchanged free of charge. Arrangements for provision of information to the public should be jointly discussed and may include the establishment and maintenance of a joint website.

Information is based on aggregated data. When exchanging information, it is important that the underlying data are available and accessible for reasons of transparency. WMO therefore recommends an open data policy as a good practice.[44] Where possible, riparian countries could aim to establish open databases following the FAIR principles.[45]

Itaipu Hydroelectric Dam on the Parana River

---

[44] Also see www.bom.gov.au/water/about/publications/document/Good-Practice-Guidelines-for-Water-Data-Management-Policy.pdf

[45] Findability, accessibility, interoperability and reuse of digital assets (FAIR) www.go-fair.org/fair-principles

Pilot monitoring exercise Armenia-Georgia, September 2020

5. SECURING FUNDING FOR MONITORING AND ASSESSMENT

In securing funding for monitoring and assessment a distinction should be made between the initial development of a system – which may involve other funding sources such as loans – and funding for maintenance and operation of an existing system. Here, it should be noted that much information can be collected at limited cost through a step-by-step approach (see Chapter 6).

Sustainable financing of monitoring systems is crucial in order to be able to identify trends and changes over time and therefore to single out the effects of policies and measures. In some contexts, joint bodies can be uniquely positioned to support transboundary activities, particularly information collection and institutional strengthening actions. Infrastructure assets such as monitoring stations are typically (although not always) developed and managed at the national level, even when the data are shared by more than one country. That said, some activities can be implemented through national and transnational actions, such as the installation and management of monitoring stations for weather information and analysis. In such a case, physical investments may be made at the national level, while a joint body provides capacity building for data collection and management, the institutional home for a database, analytical services and information dissemination.

Monitoring and assessment of water quality and quantity requires adequate resources. Therefore, those who carry out monitoring and assessment need to be able to convincingly demonstrate both the benefits of monitoring for integrated water resources management and the possible costs of not monitoring in terms of environmental degradation and other impacts. This is particularly crucial for countries in which monitoring activities still seem to be insufficiently funded.

As each basin is different, joint bodies need to identify the most suitable role for their basin in supporting the financing of the monitoring system. The costs of monitoring should be estimated before monitoring programmes begin or when major revisions are planned. If the information needs are well defined, the estimate can be detailed, with monitoring costs divided into investment costs and operation costs. Typically, the costs include the following components:

- network administration, including design and revision;
- capital costs of monitoring and sampling equipment, automatic measuring stations and data transmission systems, construction of observation boreholes or surface water sampling sites and gauging stations, transport equipment, data processing hardware and software;
- labour and other operating costs of visits to monitoring locations, sampling, field analysis of water quality variables, and field measurements of water levels and discharge characteristics;
- maintenance of monitoring stations (e.g. boreholes, automatic stations);
- development and maintenance of databases;
- labour and other operating costs of laboratory analyses;
- labour and associated operating costs of data storage and processing;
- regular training of staff, including for new instruments or systems;
- costs for quality assurance such as intercalibration exercises and general quality management;
- assessment and reporting (including joint work for transboundary waters); and
- production of outputs, including geographic information systems (GIS) or presentation software and report printing costs.

The costs associated with administration as well as assessment and reporting are largely fixed and relatively independent of the extent of the system. In contrast, the costs of other activities are strongly influenced by the number and types of sampling points, the frequency of sampling and the range of variables to be analysed. The number of sampling points can be multiplied with the frequency and variables to obtain rough cost estimates.

Because of the continuous character of monitoring, a long-term commitment to funding is crucial to ensure the sustainability of monitoring and assessment activities. Accordingly, funding should come mainly from the state budget. However, water users such as municipalities, water and waste utilities, factories, farmers and irrigators should all contribute to funding monitoring and assessment programmes. It may also be possible to raise funds by using part of the income from water abstraction fees or by invoking the polluter pays principle. Donor-funded projects concerning transboundary watercourses should be coordinated with national authorities to ensure the continuity of monitoring activities established for specific projects.

In addition, monitoring and assessment programmes for transboundary waters must be integrated into the national monitoring programmes of riparian countries. These countries should take responsibility for all costs arising on their own territory. Moreover, the riparian countries should jointly decide on funding principles and make clear agreements regarding the funding of specific joint tasks.

The report *Funding and Financing of Transboundary Water Cooperation and Basin Development*[46] provides an overview of possible sources for funding transboundary water cooperation. Such sources include public funding through (regional) taxes, user/polluter fees, management and administration fees, loans and grants, technical assistance and climate funds, although private funding sources are rare. The primary source of funding to support monitoring should be domestic budgetary resources from riparian states, but in many instances International Financial Institutions (IFIs) and projects may provide significant assistance.

Finally, when evaluating the cost of monitoring and assessment it should be recognized that the monitoring and assessment system, when well-designed, not only supplies information relevant for transboundary cooperation, but also provides valuable information for national policy-making.[47]

---

[46] UNECE, 2021. *Funding and financing of transboundary water cooperation and basin development*, available at https://unece.org/info/publications/pub/359843

[47] See, for example, www.sdg6monitoring.org/why and Timmerman, J.G., Langaas, S. *Water information: what is it good for? The use of information in transboundary water management*. Reg Environ Change 5, 177–187 (2005), available at https://link.springer.com/article/10.1007/s10113-004-0087-6

Hydropower plant in Ghana

Monitoring, assessment and data collection in Senegal

6. DEVELOPING STEP-BY-STEP APPROACHES

Summary

This chapter describes the development of step-by-step approaches to building and expanding a monitoring and assessment system – a process that entails identifying and agreeing on priorities for monitoring and assessment, and proceeding progressively from general appraisal to more precise assessments, and from labour-intensive methods to higher-technology ones. It also proposes ways to prioritize monitoring efforts in transboundary contexts and opportunities to use models in monitoring and assessment programmes.

6.1 An overview of step-by-step approaches

The monitoring and assessment of transboundary waters has multiple purposes. To make the best use of available resources and knowledge, a step-by-step approach is recommended. This entails identifying and agreeing on priorities for monitoring and assessment and proceeding progressively from general appraisal to more precise assessments and from labour-intensive methods to higher-technology ones. Such a step-by-step approach can also help specify information needs and thus focus assessment activities to ensure they are as effective as possible. A step-by-step approach is also recommended by the Guidelines on monitoring and assessment of transboundary rivers,[48] groundwaters[49] and lakes.[50]

Developing a step-by-step approach in a transboundary context may also have other implications. For example, it might mean starting with informal cooperation at an operational level and then graduating towards more formal agreements and the establishment of joint bodies as mutual trust increases. Experience also suggests setting modest objectives in the early stages – for example, regular exchanges of data and information about sampling methods and instrumentation used. This could lead to jointly agreed measurement and sampling procedures and analytical methodologies, which would pave the way to joint measurements and sampling. The eventual goal would be joint data analysis and regular joint assessments backed up by joint monitoring design.

Taking a step-by-step approach could also mean starting with data sharing for stations and sampling points close to the border and then, once this activity is well established, extending it to the whole transboundary basin or aquifer.[51] Finally, a step-by-step approach could imply starting with exchange of information on water status (quality and quantity), and once the relationship between riparian countries becomes stronger, sharing information on pressures and driving forces, evaluating the impacts on main water uses and considering possible responses – that is, applying the DPSIR framework.

Achieving the purposes and objectives of monitoring and assessment can be compared to creating a road map towards a final goal. It involves building "modules" for transboundary water monitoring and assessment, starting with tasks that can be easily accomplished in a given situation. These are followed by tasks to be carried out later once more human and financial resources, better knowledge and mutual understanding or otherwise improved conditions for transboundary cooperation have been achieved.

In countries where it is difficult to amend national legislation in the short term, a step-by-step approach could involve accepting the use of water-quality objectives or even ecologically based objectives as a basis for the monitoring and assessment work of joint bodies. These objectives could then become part of jointly agreed rules or even protocols to bilateral and multilateral agreements without the need to amend national legislation.

6.2 Prioritizing monitoring efforts

Identification of the main water functions and uses and important related issues (see Section 7.2) is crucial to determine priority information needs for water quality and quantity, and the relevant variables requiring monitoring. National surveys and land-use maps can be useful here, providing researchers with a rapid overview of possible pressures in the basin.

---

[48] UNECE Task Force on Monitoring and Assessment, 2000. *Guidelines on Monitoring and Assessment of Transboundary Rivers*, available at https://unece.org/DAM/env/water/publications/assessment/guidelines_rivers_2000_english.pdf

[49] UNECE Task Force on Monitoring and Assessment, 2000. *Guidelines on Monitoring and Assessment of Transboundary Groundwaters*, available at https://unece.org/DAM/env/water/publications/assessment/guidelinesgroundwater.pdf

[50] UNECE Working Group on Monitoring and Assessment, 2002. *Guidelines for the Monitoring and Assessment of Transboundary and International Lakes. Part A: Strategy document*, available at https://unece.org/DAM/env/water/publications/assessment/lakesstrategydoc.pdf and UNECE Working Group on Monitoring and Assessment, 2003. *Guidelines on Monitoring and Assessment of Transboundary and International Lakes. Part B: Technical guidelines*, available at https://unece.org/DAM/env/water/publications/assessment/lakestechnicaldoc.pdf

[51] An aquifer is a permeable water-bearing formation capable of yielding exploitable quantities of water.

The use of risk assessment techniques (and recording how they are applied) can help those responsible for assessments to decide which monitoring activities have the highest priority. A key concept here is "expected damage". This enables determination of what will likely go wrong in the absence of sufficient information due to lack of monitoring, and what losses will occur when less than optimal decisions are made as a result.

However, no monitoring programme can measure all variables at many sites and as frequently as would be desirable. Therefore, the monitoring design should employ risk-based approaches to select variables. For many variables, existing literature on their occurrence in the environment and particularly in freshwater systems, as well as existing information about potentially polluting activities present in the basin in question, can provide guidance for prioritization. In addition, predictions can be made as to which chemicals are most likely to reach surface water and groundwater based on their properties.

In the case of groundwater, the long-established and widely adopted approach of defining and mapping the vulnerability of aquifers to pollution can be used to prioritize monitoring. Based on the physical and chemical properties of the soil and geological materials above the water table, the potential for pollutants to be retarded and attenuated is evaluated and mapped. Where such maps exist, they can be used to help focus monitoring in areas where groundwater has important uses and where it is most vulnerable.

Risk assessment can also be used to determine whether the chosen monitoring strategy will fully meet the information needs. Statistical modelling to help optimize monitoring design (spatial density and sampling frequency) implies an element of risk analysis. For example, it can provide information on whether the resulting decreased level of information will still meet all specified information needs if either density or frequency is reduced.

6.3 Use of models in monitoring and assessment

Models (numerical, analytical or statistical) can play several roles in monitoring and assessment. They can be used to calculate water quality and quantity at certain locations, thus reducing the need for monitoring efforts. However, regular calibration with real measurements will remain necessary to calibrate the model. For assessment purposes, computer models of rivers and surrounding areas, linked to geo-referenced databases, can be used to analyse the impact of proposed measures (e.g. by simulating flow and water level variations in rivers and on flood plains during floods). Models can also be used to screen alternative management options and policies and monitoring and assessment strategies, optimize network design, and determine the potential impacts on water bodies and the associated risks to human health and ecosystems. Moreover, models play an important role in flood forecasting and early-warning systems (e.g. flood forecasting, and travel time calculations in emergency warning systems in the event of pollution from accidents and spillages).

Models should be carefully calibrated and validated with historical data to avoid unreliable results and misunderstanding of the behaviour of the basin or aquifer. Successful mathematical modelling is only possible if the approach is properly harmonized with data collection, data processing and other techniques for evaluation of the characteristics of the whole transboundary water system. It is therefore essential that the model system applied is transparent and, if possible, based on open-source software. In addition, the model structure and parameter choices must be understood and presented to the joint bodies. The preferred approach is to use several models (cloud modelling) and then present the resulting projections in joint meetings for discussion among experts. If both the conceptual model and the basic data for validation are agreed upon and reliable, then the model comparison can be carried out using the same data if the modelling software used by the riparian states is not the same.

6.4 Using pilot projects

Pilot projects are important in establishing effective and efficient monitoring and assessment programmes. They also help to initiate bilateral and multilateral cooperation, leading to institutional strengthening and capacity-building. As part of a step-by-step approach, it is desirable to implement pilot projects before setting up full monitoring and assessment systems for all riparian countries' transboundary waters. Such an approach enables the involvement of organizations with a direct or indirect stake in the use and management of transboundary waters.

A road map is an inherent and vital part of all pilot projects, stipulating the achievable objectives and clear and realistic tasks, taking into account the specific characteristics of the basin, lake or aquifer. These characteristics include the number of riparian countries and their proportions in the basin; the political, social, institutional and economic situation of the countries; and the nature of the basin. However, the commitment, resources and time needed for pilot projects should not be underestimated.

Monitoring, assessment and data collection activities in Senegal

Polluted lake, Rosia Montana, Romania

7. IMPLEMENTING MONITORING PROGRAMMES

Summary

This chapter presents the different elements of the monitoring and assessment cycle – the specification of information needs, the development of information strategies, and the data collection methodologies and management to implement monitoring programmes. It also enumerates the different data sources. The following chapters will present elements of data management and sharing, and reporting and information use.

7.1 Monitoring and assessment cycle

The monitoring and assessment of watercourses, including transboundary waters, follows a certain sequence of activities, as reflected in Figure 4.

Figure 4: Monitoring and assessment cycle

Source: UNECE, *Strategies for Monitoring and Assessment of Transboundary Rivers, Lakes and Groundwaters* (New York and Geneva, United Nations, 2006).

The outputs produced by each stage in the monitoring and assessment cycle are used in consecutive stage(s). Ideally, at the end of the cycle, the information needed for planning, decision-making and operational water management at local, national and/or transboundary levels is obtained in the form of a report or other agreed-on format. The process should also clarify the kind of information still needed to ensure better decision-making and implement other water management tasks, given that policies and/or targets may have changed in the interim. This leads to a new cycle with redefined or fine-tuned information needs, an "upgraded" information strategy and so on.

7.2 Information needs

Analysis of water management issues forms is a prerequisite to specifying information needs. Such needs relate to:

- the **water uses** (e.g. drinking water, irrigation and recreation) and **functions** (maintenance of ecosystems, protection of habitats and aquatic species) of the watercourse, which imply quality and availability requirements;

- **issues** (e.g. flooding, sedimentation, salinization, pollution, morphological alterations and damming) that hinder the proper use and functioning of the watercourse; and

- **measures** taken to address issues or improve the use or functioning of the watercourse, including environmental aspects and the protection of biodiversity.

The information needs should be clearly categorized in terms of different levels (e.g. basin-scale and local levels) and the relevant components of the DPSIR framework.

Several different activities are necessary to identify issues and priorities related to the use and protection of a transboundary river, lake, groundwater or transitional water – and their ecosystems. These include (i) identification of the uses and functions of the basin, (ii) inventories on the basis of available (and accessible) information, (iii) surveys (if information is lacking), (iv) identification of criteria and targets, and (v) evaluation of water and environmental legislation in the riparian countries in order to identify provisions of importance for monitoring and assessment (Figure 5).

Figure 5: *Analysis of water management issues*

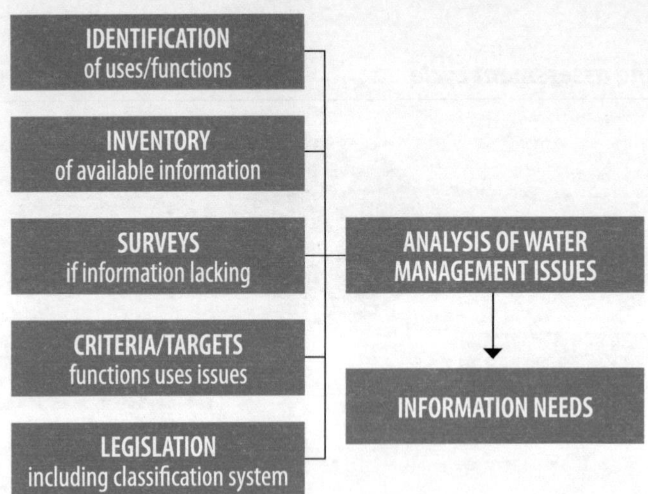

Source: UNECE, *Strategies for Monitoring and Assessment of Transboundary Rivers, Lakes and Groundwaters* (New York and Geneva, United Nations, 2006).

In order to specify information needs, information users and producers should interact closely. This means that the institutions responsible for protection and use of the transboundary watercourses, especially joint bodies, should be involved in the process of identifying and specifying information needs. Once specified, these information needs will form the basis of design criteria for the monitoring and assessment system. They should therefore be based on identified management issues and the decision-making process in basin management.[52] Throughout the process, close consideration of gender issues is essential.

Inventories of existing information will compile sources that are available, but may be incoherent and distributed among different agencies/institutions. This includes information available from historical data, licenses and similar sources in administrative databases, as well as a general screening and interpretation of all information relevant to the aspects under consideration.

Inventories should cover the main aspects of relevance to identification of the issues. Examples include water uses and water needs in the basin, run-off characteristics and the probability of flood waves and ice drifts, morphological alterations of lakes and rivers, declines in groundwater levels, droughts, water quality, state of the ecosystems as well as protected areas and aquatic species, decline in fish stocks, and sources of pollution from industries and municipal waste (especially "hot spots"). The latter should be characterized in terms of production processes, pollution composition and discharge load, land uses, and diffuse pollution sources, with a register kept of the use of fertilizers and pesticides in agriculture. Other sources of pollution may include traffic and airborne pollution (which sometimes cause acid deposition), potential sources of accidental pollution such as pipelines, and other existing point pollution sources (e.g. uncontrolled waste disposal sites). Sources may also include ore and salt deposits, which are responsible for a certain "background concentration" caused by geophysical and geochemical processes.

---

52   The following paper describes a methodology for specifying information needs: Timmerman, J. G., de Vries, S., Berendsen, M., van Dokkum, R., van de Guchte, C., Vlaanderen, N., Broek, E., & van der Horst, A. (2022). *The Information Strategy Model: a framework for developing a monitoring strategy for national policy making and SDG6 reporting.* Water International, 47(1), 55-72, available at www.tandfonline.com/doi/full/10.1080/0 2508060.2021.1973856

Surveys will be needed if the inventory does not provide sufficient data. Water quality surveys are intended to give a first insight into the structure and functioning of the aquatic ecosystem and the occurrence of pollution and toxic effects in the water. Investigating the qualitative and quantitative structure of the biocoenose[53] concerned makes it possible to assess the ecological status of a river, lake or transitional water. This may involve chemical screening of surface water, groundwater, sediments and effluents at hot spots and key locations. Additionally, analysis of specific target compounds that might be expected can be performed, according to the inventory, with toxic effects in surface water, sediments and effluents investigated at these locations. Surveys of water usage may also be required.

Water balances[54] or water accounts[55] should be drawn up for (parts of) a basin, especially lakes and aquifers, when and where the careful sharing of available water resources for different water uses is of particular importance. Water management balances compare water resources with water uses, water consumption and ecological water demand. In addition to undisturbed river run-off, a water management balance includes, for instance, intakes from and discharges to the river by municipalities, industries, irrigation, drainage and fish farming; diversions from and into the river; reservoir storage and release; discharge of groundwater resources into the river, mine dewatering and so on.

Healthy ecosystems are essential for resilience and sustainable development, and provide essential products and benefits. Water quality and quantity are of particular importance for ecosystems and should be included in information needs.[56] These include not only aquatic ecosystems, where a flow regime (e.g. ecological flow) must be maintained, but also ecosystems that are dependent on groundwater, where certain levels may be critical for maintaining ecosystem health.

When specifying information needs it is important to bear in mind the design of the monitoring and assessment programme. At a minimum, the specified information should determine:

- the appropriate variables for monitoring;
- the criteria for assessment (e.g. indicators, early warning criteria for floods, droughts or accidental pollution);
- protected areas and other vulnerable and valuable environments that need to be considered;
- specified requirements for reporting and presenting information (e.g. presentation in maps, GIS, degree of aggregation);
- the relevant accuracy for each monitoring variable;
- the degree of data reliability; and
- the specified response time (i.e. the period of time within which the information is needed) for forecasts or early warning systems (e.g. minutes/hours), trend detection (e.g. number of weeks/months/years after sampling) and other tasks.

The relevant accuracy and degree of data reliability are decisive factors in the selection of monitoring sites, the determination of monitoring frequencies, and the choice of laboratory technology and methodologies for data management.

Finally, information needs should be prioritized. When the same information need arises from a variety of water management problems, it should be accorded a high priority, as gathering this information once makes it possible to address a variety of issues.

7.3 Information strategy

After information needs have been identified, specified and prioritized, an information strategy should be developed. This strategy defines the most practical way of gathering data from various sources including the monitoring system, expert judgments, statistical publications, open data sources, remote sensing, citizen science, Indigenous and local knowledge, and the document libraries of various institutions (see Section 7.4). The information strategy should culminate in a monitoring plan and a separate plan for gathering data from a range of sources.

---

[53] A biocenosis describes the interacting organisms living together in a habitat.

[54] A water balance is used to describe the flow of water in and out of a system. The water balance tracks water input and output as well as the different forms it can take as liquid, solid (snow and ice) and gas (evaporation).

[55] Water accounting is defined as the systematic acquisition, analysis and communication of information relating to stocks, flows and fluxes of water (from source to sinks) in natural, disturbed or heavily engineered environments (www.wateraccounting.org).

[56] See the training manual Global Water Partnership, 2021. *Integrating Data to Improve the Protection and Restoration of Freshwater Ecosystems*, available at https://www.gwp.org/globalassets/global/activities/act-on-sdg6/microsite/661-page/2021_capnet_gwp_training-manual_freshwater_ecosystems_compiled-1.pdf

The information strategy needs to be adapted over time as water management develops, targets are attained or policies change. However, ensuring continuity is essential in order to produce time series that allow for the detection of significant and reliable trends. In general, environmental monitoring programmes should be understood as a long-term commitment.

7.4 Monitoring/data collection

The main monitoring objective for rivers, lakes and groundwaters as well as effluents is to generate information to be used both in national and transboundary contexts. This information may be used for:

- assessing the actual status of water resources;
- detecting possible long-term trends in water levels and discharge or pollutant concentrations;
- providing for hydrological forecasts;
- assessing pollution loads from point and non-point sources;
- testing for compliance with permits for water withdrawal or discharges of wastewater, and establishing taxes, fines and sanctions;
- verifying the effectiveness of policy measures;
- contributing to reporting on the state of the environment;
- providing early warning to protect intended water uses in the event of flooding, drought or accidental pollution;
- recognizing and understanding processes in water and water-related ecosystems (e.g. flow regime, erosion patterns, morphological alterations and damming, hydrobiological processes, natural or background pollution of water bodies);
- enabling assessment of imminent or possible health risks, and supporting predictions of long-term processes which may have health-relevant outcomes; and
- revising, if appropriate, current monitoring and assessment activities, including the existing monitoring system.

Each of these objectives may require specific measurement devices or sampling procedures.

The most resource- and labour-intensive stage in monitoring is the phase that includes sampling, in situ physicochemical analysis, hydrobiological and water-quantity-related measurements, and laboratory analysis. This phase also entails high risks in producing reliable and accurate data. To support such analyses and to facilitate consistency across borders, it is essential that the data are compatible, comparable and of controlled quality.[57] Therefore, it is important to employ qualified and experienced personnel and comply with internationally agreed guidelines and standards.

Standards are essential to secure the compatibility, comparability, interoperability and quality of data and information. A standard is a document – preferably based on an internationally agreed standard (e.g. International Organization for Standardization (ISO) – that provides requirements, specifications, guidelines or characteristics that can be used consistently to ensure that materials, products, processes and services are fit for their purpose. The aims and benefits of standards are to:

- improve quality and trust,
- enable sharing of data,
- increase comparability of measurements, and
- improve understanding of uncertainty.

The process thus becomes outcome and information oriented. In terms of characteristics, standards should be performance focused, precise, complete, unambiguous, state of the art, and comprehensible to qualified individuals who have not participated in their development.

Standards are of particular importance in the transboundary context, it is crucial that a standard-setting organization be appointed to oversee the monitoring and data collection process, and that procedures are in place to support

---

[57] UNECE Task Force on Monitoring and Assessment, 1996. *Quality Assurance*, available at https://unece.org/DAM/env/water/publications/documents/quality_assurance.pdf

this task.[58] Next to standards, laboratory proficiency testing and intercalibration exercises[59] are needed to ensure comparability.

Additional data collection will be needed to make an integrated assessment of boundary conditions and the broader social and environmental context. This includes human activities that may influence the state of the water resource and the ecosystem, as well as policies and plans.

7.5 Different data sources

The integrated assessment will require data collection from a range of sources including expert judgements, statistical publications, open data sources, policies and plans in addition to monitoring data. As different organizations can provide different types of data and information, cooperation will be essential.

In addition to more traditional monitoring through in-situ visits and sampling, recent technological developments enable monitoring approaches that may reduce the workload and resources needed or increase the associated amount or detail of information. While some technologies have not matured sufficiently, others provide good practice examples. Below is a non-exhaustive list of developments. Depending on the specific basin and monitoring needs, specific technologies can be selected (e.g. remote sensing may not be applicable for groundwater monitoring).

7.5.1 Remote sensing and Geographical Information Systems (GIS)

Remote sensing, and satellite imaging in particular, has developed significantly in recent years.[60] Advantages include the fact that remote sensing covers a wide area, eliminating the need for in-situ visits, except for ground-truthing purposes. Disadvantages include the possibility that clouds may hinder satellite imaging, the relatively low level of detail (resolution), the fact that imaging is mostly limited to the water surface and that temporal coverage may be limited. Applications of the technology for water quality, for instance, are also limited.

Combined with GIS, satellite images can be used to provide good information on land use, vegetation and soil moisture, among others. Models in combination with GIS provide opportunities to identify hot spots and show geographical relationships. In addition, significant amounts of open access data are available for GIS, while GIS can form a good basis for the sharing and exchange of data.

7.5.2 Self-monitoring

Emission registration by companies (self-monitoring) provides a good source of information, particularly for water quality purposes. This process entails obliging companies to report on their emissions, discharges and losses to air, water and soil. Such an obligation can be part of the operating license. However, regular checking that the reports reflect actual emissions is necessary, and is usually done through inspections.

Such a registration system covers point source pollution. For diffuse pollution, various methods exist that enable calculating pollution loads from various sources, including agriculture and road and rail transport. With this information, estimates can be made of the various sources of pollution. This in turn provides information on where measures can be effective.

7.5.3 Citizen science

Citizen science[61] is a process by which everyday people actively contribute to research and monitoring. There is a long tradition in hydrological monitoring of using local people to make observations and then feeding the data into official hydrological databases. Regarding water quality, citizens can be provided with simple testing kits with which they can monitor water quality. Additionally, mobile phones have significantly simplified and accelerated the reporting phase for this type of monitoring. However, application of citizen science in monitoring requires a long-term commitment on the part of citizens as well as sufficient training by professionals to ensure the quality and credibility of data. Gender aspects are also relevant when developing citizen science platforms and determining the

---

[58] See the Manual on the WMO Information System (WMO-No. 1060): Annex VII to the WMO Technical Regulations, 2021 update, which sets out standard and recommended practices and procedures, available at https://library.wmo.int/index.php?lvl=notice_display&id=9254#.YgPMly-iFaR

[59] See, for example, ISO/IEC 17025:2017 General requirements for the competence of testing and calibration laboratories (www.iso.org/standard/66912.html)

[60] See, for example, World Bank, 2019. *New Avenues for Remote Sensing Applications for Water Management: A Range of Applications and the Lessons Learned from Implementation*, available at https://openknowledge.worldbank.org/handle/10986/32105

[61] See, for example, https://citizenscience.org

set of observations to be collected and consequent information flows, in order to ensure equal opportunities for reporting, access and reachability of the information gathered.

7.5.4 Drones

Drones are vehicles that can be operated from a distance. A floating drone can, for instance, be programmed to make transects in a lake and take a sample or measurements at regular intervals. This is in general cheaper and often more precise than a boat manned by individuals performing sampling. When sampling must be performed at different depths, a submarine drone can be used. These devices are mostly still under development but are expected to develop quickly. The use of drones requires relevant permissions/licenses for operations, and specific attention should be paid to sharing and open access to data and videos in the joint basins.

7.5.5 Sensors

The increasing availability of (automatic) sensors able to measure certain variables is gradually leading to the displacement of chemical analysis. These sensors enable the installation of automated monitoring stations that collect data continuously or at regular intervals. Communication technology then allows for remote online collection of the data. Sensors can also be installed on ferry boats to collect regular transects, or on drones. However, the sensors need regular maintenance and cleaning.

The amount of data collected using automated sensors is increasing exponentially. Therefore, it is important to ensure that data transfer and quality control procedures for large amounts of data are in place and included in the cost of monitoring.

7.5.6 Environmental DNA

Environmental DNA (eDNA) refers to DNA from organisms that can be found in the aquatic environment and sampled and monitored. Using eDNA may allow for rapid, cost-effective and standardized collection of data about species distribution and relative abundance. However, these analyses are not often used in operational monitoring, and require specific protocols, instrumentation, and trained staff if they are to be used in environmental monitoring and assessment.

Monitoring, assessment and data collection activities in Senegal

Groundwater purification area, The Netherlands

8. MANAGING AND SHARING DATA AND MAKING ASSESSMENTS

Summary

Managing and sharing data is a crucial component of assessments. Data management should be developed at the national level and then disseminated to cross-boundary stakeholders. This chapter describes the different processes involved in managing, storing, analysing and sharing data. It also provides an overview of assessment methodologies and the potential challenges and benefits to sharing data in transboundary contexts.

8.1 Data management

It is of utmost importance that policy-makers and planners have a clear understanding of the various steps involved in data management. Such understanding will facilitate data sharing among the institutions undertaking monitoring and assessment, including joint bodies. It may be useful to begin inter-institutional cooperation and data sharing at the national level before implementation at the transboundary level.[62] To safeguard the effective future usage of any data collected, the following guidelines should be considered.

8.1.1 Adopting strategies and rules for data management and sharing

Rules and basic principles should be established for organizing data management and sharing. The EU-wide Shared Environmental Information System (SEIS) provides an example of such rules and principles for the integrated collection, exchange and use of environmental data and information across Europe (Box 1).

Box 1: Principles of the European Shared Environmental Information System (SEIS)

The main principles of the European Shared Environmental Information System (SEIS) are as follows:

- Information should be managed as closely as possible to its source.
- Information should be collected once and then shared with others for multiple purposes.
- Information should be readily available to public authorities and enable them to easily fulfil their legal reporting obligations.
- Information should be readily accessible to end-users, primarily public authorities at all levels from local to European, to enable them to assess in a timely fashion the state of the environment and the effectiveness of their policies, and to design new ones.
- Information should also be accessible to end-users, both public authorities and citizens, to enable them to make comparisons at the appropriate geographical scale (e.g. countries, cities and catchments areas) and to participate meaningfully in the development and implementation of environmental policy.
- Information should be fully available to the general public, after due consideration of the appropriate level of aggregation and subject to appropriate confidentiality constraints, and at the national level in the relevant national language(s).
- Information sharing and processing should be supported through common, free open standards.

Source: INBO, *Handbook on Water Information Systems* (INBO, 2018).

In the application of these basic principles – and considering that data should be managed as close to the producer as possible – there is a strong case for data producers to establish their own database/information system (or have access to an external one) in order to manage and process their own data and ensure quality control.

To facilitate the aggregation and processing of data to generate information, links can be developed between producers and data managers, who generally act:

- at the national level through national thematic databases/information systems for the production of national information,

---

[62] Also see Bureau of Meteorology (Australia) and WMO, 2017. *Good practice guidelines for water data management policy: World Water Data Initiative*, available at https://library.wmo.int/index.php?lvl=notice_display&id=20165#.Y3mbqC-QmL1

- at the basin level for basin water management, and
- at the local level for local water management.

8.1.2 Developing a data dictionary

To facilitate the comparability of data, clear agreements should be made between neighbouring countries on the definition, coding and formats of collected data and supporting information. In addition, collected data should include "metadata" such as the date, location, measuring depth and measured values. Metadata should also include the assumptions and limitations affecting the creation of the data. Metadata thus allow a producer to describe a dataset fully so that users can understand the assumptions and limitations and evaluate its applicability for the intended use.[63] The adoption of internationally standardized formats for data exchange simplify technical implementation of the system and enable exchange at a wider scale. This entire process can be encapsulated in a data dictionary explaining the coding and defining terms, which should be prepared and agreed upon.

8.1.3 Data quality control

Data quality control is an intrinsic part of data management and includes regular checking and controlling newly collected data according to procedures defined and documented in the data dictionary. This approach allows for the detection of outliers, missing values and other obvious errors. Available computer programmes can perform various control functions, but expert judgment and local knowledge of water systems are indispensable for data validation. It is especially important that the data dictionary specify the level of quality control associated with each dataset. This allows potential users to clarify in advance whether the quality of the data is sufficient for the intended uses.

8.1.4 Data storage

To ensure availability for future use, data should be stored properly in databases, with sufficient supporting information to enable interpretation, comparison, processing (conversions, etc.) and reporting. The database should include safeguards against the entry of data without metadata. Furthermore, the original raw data shall be saved, separate from the refined data.

Databases can either be national or shared between riparian countries. In all cases, it is essential to ensure information availability for all riparian countries.

8.1.5 Data analysis and interpretation

The conversion of data into information involves analysis and interpretation. Data analysis should be embedded in a data analysis protocol (DAP) which clearly describes how the data should be analysed and interpreted and what should be done in cases of missing data, outliers, non-normality and serial correlation.

Data analysis can be a statistical operation or a set of operations using generic software packages. Statistical techniques may be used to detect trends and trend reversals and test for compliance with standards, although the use of tailor-made software adaptations may be desirable. The DAP should therefore include procedures for processing the monitoring data to meet specific interpretation needs (e.g. calculations based on individual measurements or yearly averages, single sites or averages for the whole water body) based on the type of water system.

The DAP should also integrate reporting formats for the resulting information. In concrete terms, the DAP will need to specify the format of the report, the frequency of publication, the intended audience, distribution procedures, and the types of conclusions to be drawn and represented.

8.2 Assessment methodology

The assessment methodology will determine or at least influence the design of the monitoring programme. Therefore, it should be drawn up in parallel with the analysis of information needs and design of the monitoring programme, with a focus on the legislative context, policies and issues.

Given the purposes of assessments, a simple way of using monitoring results is to focus on certain key variables and indicators. For example, in cases where binding water protection targets for certain pollutants, like pesticides, have been expressed by numerical norms or standards, establishing the state of watercourses in these terms is a straightforward task and can be done at a very early stage. Another simple yet informative method of assessment

---

[63] ISO norm 19115:2003(E).

is to prepare maps of the distribution of monitored variables for certain larger water areas. Such an assessment can be particularly appealing and understandable for laypeople. Where standards and norms differ between riparian countries, normative targets can be used to compare between the countries.

In monitoring programmes where large amounts of different data are collected continuously over several years, statistical methods are needed to effectively summarize the results. In particular, different types of trend calculations are being used to assess monitoring data. When interpreting trends in water quality, particular attention should be paid to water quantity data, since hydrology strongly affects water quality. For example, flow normalization is regularly used to assess and compare pollutant loads.

The use of water classification systems to assess watercourses is very common. Some of these systems are based on physicochemical variables, but biological approaches (e.g. ecological classification under the WFD) are also used. For transboundary water assessments, whether based on classification systems or other assessment methods, it is important to initially strive for comparability of results rather than unification of methods and standards, as the latter can be a very long process.

Since groundwater systems are three-dimensional, often complex environments, with limited observation points (e.g. springs, wells), their assessment usually requires expensive and long-term efforts. For instance, it is not unusual to find several aquifers on top of each other exhibiting different and varying flow directions and rates, or groundwater quality. Sufficient resources should be mobilized to sustain these groundwater assessment activities in the long run.

8.3 Data sharing

At the transboundary level, sharing of information and data between countries is often difficult for political/structural reasons, particularly when there is no agreement or protocol between the countries on data sharing, and technical reasons, notably difficulties related to information collection, harmonization of data formats, definitions, analysis methods, frequency of data collection, density of monitoring networks and data processing. National authorities may also be reluctant to provide neighbouring countries with information that they consider strategic (e.g. the economic value of water used for hydropower, agricultural irrigation and navigation).

To ensure smooth data sharing, both transboundary and national institutions have to resolve certain questions:

- What is the optimum way to organize the production of new datasets and the enhancement of existing ones, in order to generate information and useful services for decision-making purposes, and to inform partners and the public?

- What datasets already exist, in what form, and how can they be accessed and integrated in a flexible and efficient manner? Additionally, how can they be preserved from deterioration and loss?

- What are the best ways to manage the multiplicity of data producers and available formats as well as the issue of comparing datasets that are often incomplete, dispersed and of variable quality?

- What legislative/institutional frameworks exist to organize the sharing of data among partners as well as the processing and dissemination of the results?

Taking into consideration that data management is primarily a tool to support water policy, its organization at the transboundary level will depend to a large extent on the type of existing joint bodies and the level of cooperation defined in the provisions of agreements between the countries. Here, a joint body with an operative secretariat may be able to allocate human and financial resources to improve data sharing; to organize transboundary data processing and information dissemination; to support/complete the data production processes existing at national level; and to develop and manage the information system, where not based on national systems.

However, when no secretariat with specific resources exists, it will be necessary to draw on the resources of national organizations to support these processes, or to rely on external resources. Where data sharing has been initiated as part of a project, it is important to consider the issue of sustainability of the processes.

In any case, as the majority of the data used for transboundary water resources management is provided by national organizations, the transboundary information system should ideally be built to rely on national information systems with (direct) access to datasets made available by national partners. This implies a need to reinforce national capacities in data management and to develop capacities to exchange comparable data and ensure interoperability with the information systems of partners, using a common language (concepts/referential dataset) and common procedures. Moreover, formats for the exchange of data should be defined and agreed upon by the users (see section 8.1.2 on the data dictionary).

In some situations, for example, where a large number of countries share the basin, the relevant joint body may consider the establishment of a common platform and common procedures to facilitate data storage and exchange. The WMO Unified Policy for the International Exchange of Earth System Data[64] and the guidelines for data sharing developed by EUROWATERNET[65] may be suitable to support such activities.

---

[64] https://public.wmo.int/en/our-mandate/what-we-do/observations/Unified-WMO-Data-Policy-Resolution

[65] http://dd.eionet.eu.int/index.jsp

Biological sampling in Serbia during Joint Danube Survey 4

Okavango River delta in Botswana

9. REPORTING AND USING INFORMATION

Summary

This chapter describes the crucial role of reporting in the monitoring and assessment cycle. Information about water resources contributes to environmental reporting and may inform planning relevant for water-using sectors. Reporting includes the production of reports and other modes of disseminating information, such as application programme interface newsletters (API) and online applications. These are all instrumental in ensuring the dissemination of interpreted data, and the use of information produced to inform management decisions between targeted stakeholders. There are several reporting obligations which vary according to the type of international convention or transboundary agreement.

9.1 Information dissemination

Reporting is another essential step in the monitoring and assessment cycle, and is vital to disseminate the resulting information. Reporting plays a key role in decision-making for water management and the further development of monitoring and assessment programmes. Reporting is not limited to producing a report but also entails all types of dissemination including API-generated[66] newsletters and online applications.

Information about water resources also contributes to environmental reporting and may inform planning relevant for water-using sectors. Information dissemination should take place on a regular basis, and the interpreted data should be made available in an easily accessible and understandable manner tailored to the audience being addressed. This is best done by providing the information from the monitoring in the context of legislation, policies and/or specific issues on which the information is collected, as specified through the information needs.

The same information should be ready to be used for a variety of purposes, including different reporting obligations, and by a variety of users. Any environmental information system should therefore serve a range of different purposes and not be designed for any one single purpose.

9.2 Reporting obligations

Environmental information should be public, according to principle 10 of the Rio Declaration, the Aarhus Convention and the Escazú Agreement. Reporting of environmental information plays an especially important role in increasing public awareness of water problems and promoting public participation in water management.

In order to cost-effectively fulfil reporting requirements laid down in national water management legislation, applicable transboundary agreements as well as relevant decisions taken in international forums, an inventory should be made of national and international reporting obligations. For transboundary river, lake and aquifer basins, reporting under the Water Convention and on SDG indicator 6.5.2 is of particular relevance.

9.3 Reporting formats and audiences

The level of detail included in reports and the frequency of compilation also depend on the target audience. The content of reports should be targeted to the needs of an audience which includes international bodies, management and scientific institutions, national administrations and the public. Depending on the needs of the target group, the report may contain aggregated data (e.g. indicators) and/or detailed information in tables, statistically processed data, graphs and geographically presented information.

Public authorities, including joint bodies, usually request information in a specific format and frequency, which are defined in reporting protocols or schemes. Such reports are most often presented in writing to ensure unambiguous understanding of the results. In addition, public authorities may present ad hoc requests for information which are not predefined in reporting protocols but are related to specific current topics in water management. This kind of reporting has to meet strict requirements concerning response time and flexibility. Reporting or information on water resources may also be necessary in relation to environmental, health or economic development, where water-using sectors are involved.

Dissemination of information to the public can involve the development of shortened (online) versions of regular reports produced in easy-to-understand language. Relevant guidance is provided in the Aarhus Convention and the Escazú Agreement, among others. Specific attention should be paid to the accessibility of information, notably

---

[66] An Application Programming Interface (API) is a software interface that allows two applications to interact with each other without any user intervention.

by children, youth, elderly people, women, Indigenous people and minorities who may experience difficulties in accessing such information.

A state-of-the-environment report should provide concise information for decision-making in water management. These reports typically provide information on the state and functions of the water body, describe existing problems and the pressure they put on the water body, and give insights into the impacts of corrective measures. Their decision-making value is strongly increased when accompanied by visualization tools and indicators, particularly if elements of the DPSIR framework are reported.

The form of a joint report for the purposes of water management in transboundary basins should be agreed upon in detail by the riparian countries. Harmonization of reporting is strongly encouraged, with joint reporting naturally requiring a high level of data comparability. The reports should highlight the links between policy measures and the status of the water body of concern. Periodic assessments under the Water Convention, covering all transboundary water basins, are also recommended to encourage the evaluation of progress made under the Convention, stimulate the commitment of the members involved and make results available to the public.

API newsletters and online applications provide powerful tools for sharing and communicating information and can be used to inform and involve the public. Updated recommendations on the more effective use of electronic information tools[67] under the Aarhus Convention provide useful guidance in this respect. Some authorities have been cautious in presenting environmental information and data to the public because of the risk of misinterpretation of information by laypeople. However, involving non-governmental organizations and the public in transboundary water management promotes awareness, helps to identify omissions and mistakes, and stimulates more sustainable cooperation between countries.

9.4 Information use

The information produced must be used and should contribute to management decisions. Therefore, information products in their various forms need to be made relevant, accessible and attractive to users. These products should convey the messages that the information users really need.

The information product should be based on information needs as specified above. In particular, the information should be clearly linked to the relevant components of the DPSIR framework. While much of the information derived from a monitoring programme reflects the status of transboundary waters, it is important to include interpretation and assessment in relation to drivers and pressures and how these are changing over time, especially in relation to impacts on the health of water users, for example. Water managers will also require information products related to specific responses – for instance, the effectiveness of measures to protect or restore a basin. The information product should consequently address the full range of the DPSIR framework, in order to enable decision-making about future actions and measures.

9.5 Revision of the monitoring and assessment system

As monitoring and assessment is a cycle, the use of information should feed back into the design of the monitoring programme, leading potentially to revision and improvements, as well as to the review of and possibly changes in information needs and consequent priorities for monitoring and assessment. These include reviewing the most effective use of available funding and ensuring the inclusion of all relevant stakeholders in the information cycle. While the monitoring and assessment programme requires stability and continuity to meet information needs, the specific activities that make up the cycle should be sufficiently flexible to suit changing drivers and pressures, new legal requirements and obligations, new information needs of society and otherwise changing conditions. The monitoring and assessment cycle should, therefore, be seen as an inclusive, continuously evolving and gradually improving spiral.

---

[67]   See *Updated recommendations on the more effective use of electronic information tools* (ECE/MP.PP/2021/2/Add.2), available at https://unece.org/environment/documents/2022/02/updated-recommendations-more-effective-use-electronic-information

Joint Danube Survey 4, measuring physicochemical parameters

Monitoring, assessment and data collection activities in Senegal

ANNEXES

Annex 1. Specific aspects of groundwater monitoring

This annex represents an updated synthesis of information published in the *Guidelines on Monitoring and Assessment of Transboundary Groundwater*.[68] Specifics on groundwater monitoring can be found in these guidelines, particularly on pages 41–42 and in the box "Nine basic rules for a successful monitoring programme".

Characteristics

A distinguishing feature of groundwater compared to surface water is slow movement (long residence times). This increases the potential for changes in quality due to interactions between the water and the surrounding aquifer material. Such interactions cause the natural hydro-geochemistry to evolve as the infiltrating groundwater moves down. Once groundwaters are polluted as a result of human activity, they may remain so for many years, and it is difficult to intervene effectively in this process.

Groundwater can be replenished by precipitation or from surface water bodies, but some aquifers have little or no interaction with the surface and the water is not renewable. In addition, groundwater can have a high spatial variability, especially in heterogeneous hydrogeological settings.

Groundwater flow can be intergranular or through fractures or karst openings. Groundwater flow will be much more rapid but variable and difficult to estimate through intensely fractured or karstic rocks.

Important variables

To be able to detect and quantify trends of groundwater under natural and anthropogenic drivers, the "baseline" quality and quantity of groundwater, with its spatial (including depth) and time variations, must be assessed. This usually comes with the measurement of groundwater levels, physical parameters and chemical composition of groundwater. Groundwater abstraction is also an important variable to be monitored. Recharge and discharge areas need to be determined and activities that might affect the quantity or quality of groundwater need to be understood. Moreover, the interaction between surface and groundwaters, as well as between groundwaters, needs to be clear. Accordingly, in order to characterize groundwater flow systems, information on geology, geophysics and hydrogeology in the transboundary area, and the aquifer system in particular, is important. The dynamics of the groundwater flow system, including natural and anthropogenic variations and changes, is key information.

Climate change also affects groundwater levels. For this reason, it is recommended to monitor groundwater table levels and to determine the quantities extracted from groundwater wells, in order to maintain the aquifer within safe limits and to prevent a continuous decline in water level.

Frequencies

The monitoring of groundwater quantity differs from the monitoring of groundwater quality. Groundwater quantity monitoring is usually conducted via a monitoring network, where the same wells are monitored at frequent intervals. Groundwater quality monitoring is often done through surveys. As groundwater flow is generally slow, frequency of monitoring must be adjusted to the groundwater flow system.

Groundwater quality and quantity vary in space and time, but on different spatial and temporal scales to surface waters. This variability is made more complex by the interactions referred to above as well as with the geological media. The choice of type and location of observation points and depths is usually governed by the specific representativeness of the observation points in the aquifer and the possibility of determining the spatial trend in the groundwater levels and quality at the required scale.

Locations

Aquifers are three-dimensional, sometimes complex environments. Monitoring data must therefore be collected in different places and at different depths.

---

[68] UNECE Task Force on Monitoring and Assessment, 2000. *Guidelines on Monitoring and Assessment of Transboundary Groundwaters*, available at https://unece.org/DAM/env/water/publications/assessment/guidelinesgroundwater.pdf

Monitoring sites for groundwater level observation can be wells, springs or boreholes. The sites or observation points of a network should be representative of the delineation of the relevant groundwater flow systems and the extent of aquifers, aquitards[69] and aquicludes[70]. Accessibility of monitoring locations can be a limiting factor.

Knowledge of the groundwater flow system involves in particular the locations of groundwater recharge and discharge zones, and the way groundwater flows through the subsurface (Figure 6). This can be challenging in the case of aquifer systems, where several aquifers and confining layers occur. Furthermore, activities on one side of the border might adversely affect the quality of quantity of groundwater on the other side (Figure 7).

The desirable or target density of a network will depend on the complexity of the groundwater flow system as well as the purpose of the monitoring. Complex groundwater flow systems will require a denser network of monitoring sites. For aquifers affected by intensive exploitation and/or other anthropogenic activities (e.g. industry, intensive agriculture, landfills, abandoned municipal or industrial sites, etc.), the network density should be higher. As a general rule, factors such as aquifer characteristics, vulnerability, groundwater exploitation, water use and land use, and population served with groundwater, would be considered when developing or updating a groundwater monitoring network.

Figure 6: Transboundary groundwater flow systems

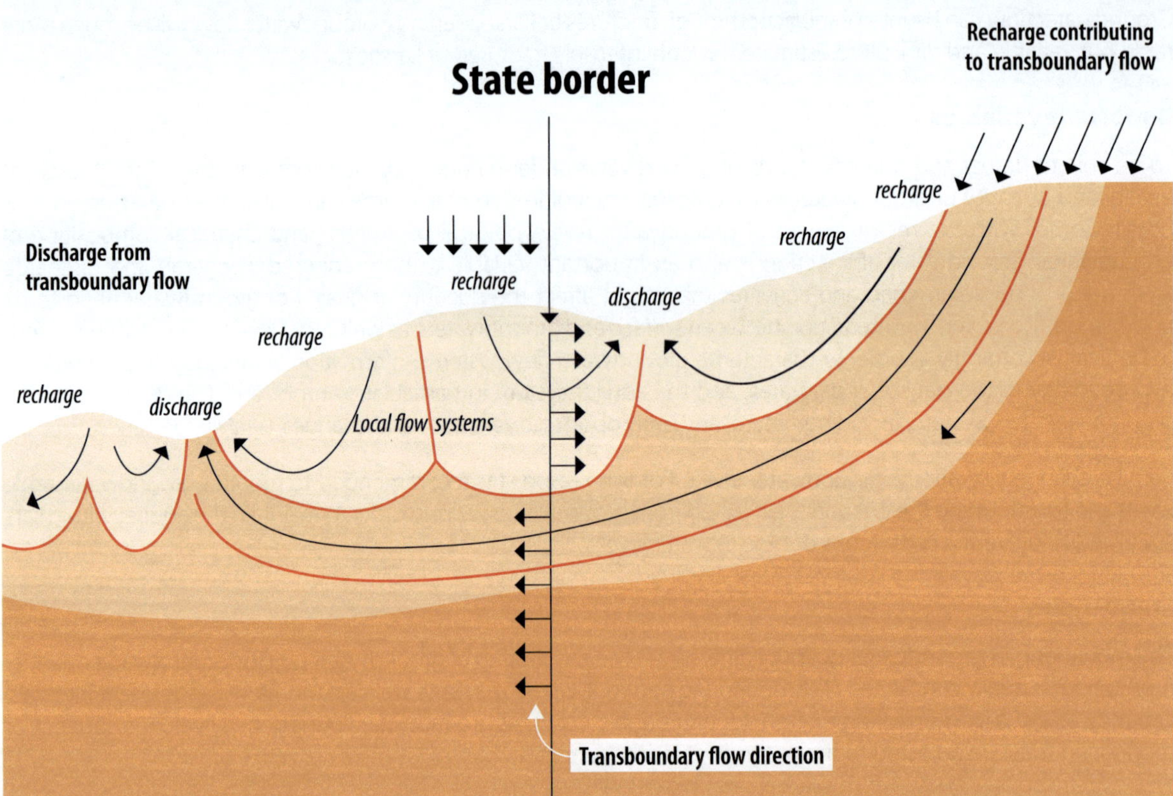

Source: UNECE Task Force on Monitoring and Assessment, 2000. *Guidelines on Monitoring and Assessment of Transboundary Groundwaters.*

---

[69] An aquitard is a formation with a relative low permeability with respect to surrounding formations.

[70] An aquiclude is an impermeable area underlying or overlying an aquifer.

Figure 7: *Effect of a transboundary aquitard on groundwater flow*

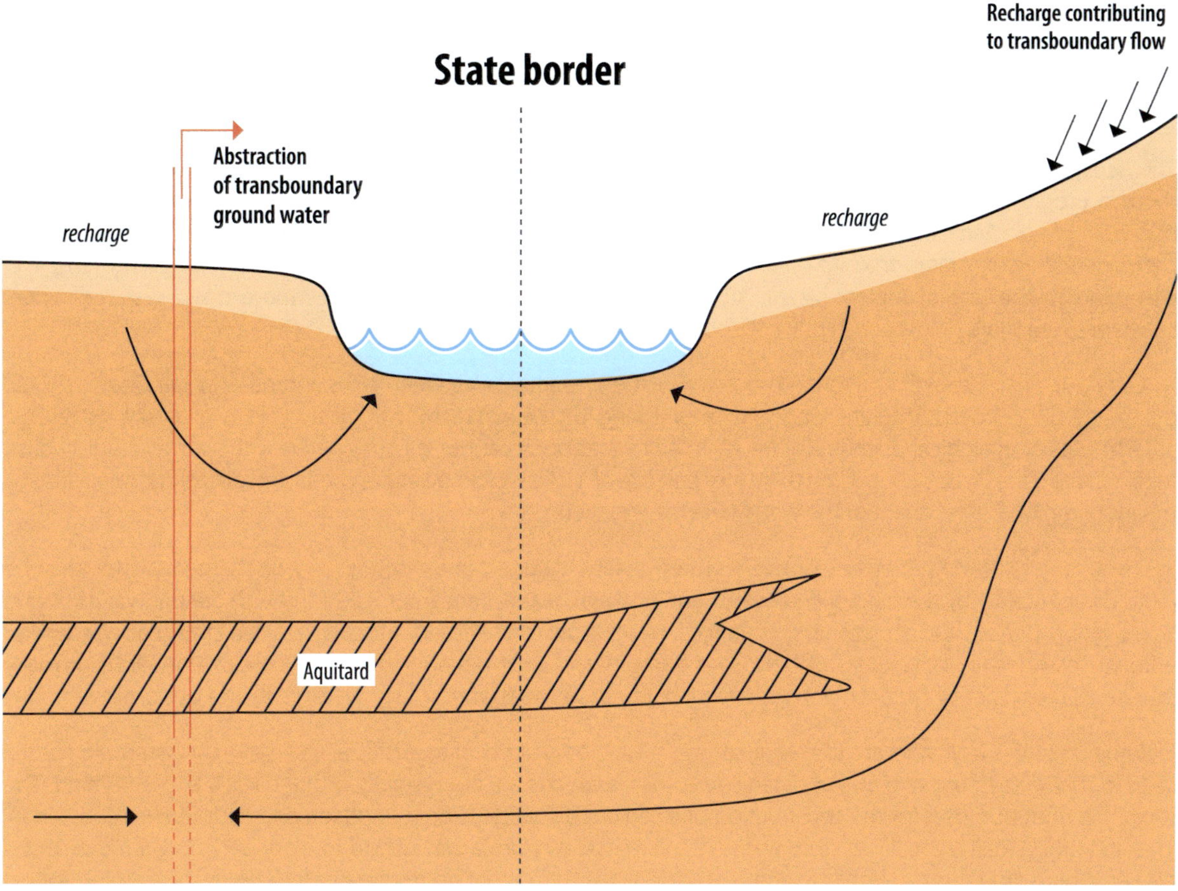

Source: UNECE Task Force on Monitoring and Assessment, 2000. *Guidelines on Monitoring and Assessment of Transboundary Groundwaters.*

Observation well in Finland

Annex 2. Specific aspects of lake monitoring

This annex represents an updated synthesis of information from the *Guidelines on Monitoring and Assessment of Transboundary and International Lakes.*[71]

Characteristics

The ecosystems of lakes differ from those of rivers in many respects, including their hydrological circumstances, thermal properties, production/decomposition relations, sedimentation rates and composition, and the stability of certain phenomena. Lakes are almost closed systems. Substances introduced into a lake therefore may become permanently incorporated into its cyclical processes, with only a proportion of the total load being removed, according to the rate of replenishment and sedimentation. Conversely, rivers are more open systems, where substances are more or less constantly transported downstream.

To carry out a reliable lake monitoring programme, the interactions between lakes and other water bodies should be clearly understood. Accurate long-term monitoring of the entire hydrological cycle is essential, as reliable assessments of ecological or chemical trends in any water body cannot be conducted without hydrological data. Most crucially, the factors controlling the water balance of a lake should be either measured directly or calculated by means of regional assessment or the water balance equation.

In many rapidly flowing stretches of rivers, water quality is quite homogenous, and discharged wastes may be diluted very quickly by the natural river water. But in lakes, wastewater can proceed through deeper waters during stratification periods for considerable distances without any real mixing. Indeed, heavier industrial wastewater effluents can destroy large areas of bottom sediments and their biota in this way. The concentrations of many pollutants may differ by factors of tens or even a hundred between the surface water level and bottom levels.

In some countries reservoirs are the most common type of water bodies. Artificial reservoirs can resemble natural lakes in many ways, but one crucial difference is that reservoirs are always built with a particular use in mind. The most common purposes for the construction of reservoirs are water supply, irrigation and hydropower generation. The underlying idea is usually to store water, delaying its flow from a wet period to a dry period, when demand is higher. Where droughts can last for many years, some reservoirs store up to three or four times the average annual flow. Exchange of information on the operation of reservoirs is particularly important for transboundary cooperation.

The dominant biological process in rivers is the decomposition of organic matter, while primary production is much less important. In contrast, in deeper lakes with clear thermal stratification, the dominant biological phenomenon in the upper water layer during summertime is primary production. In the deeper layer, primary production cannot normally be detected, and the dominant process is the decomposition of organic matter by bacteria.

Sedimentation is also a very important process in lakes, and plays a dominant role in nutrient cycles, and thus also in the eutrophication process. Sedimentation zones must be identified before monitoring programmes are implemented.

Important variables

From the point of view of the lake water balance, the key hydrological variables are typically regional precipitation, lake inflow, lake water level, lake evaporation and lake outflow. Snow cover in mountainous areas and groundwater storage are also important factors in many cases. Important physical hydrological phenomena such as sediment transport, erosion, water temperature and ice phenomena can also affect chemical and biological processes in lakes. In addition, residence times have a considerable effect on both eutrophication and the rate of recovery of polluted lakes.

The total volumes and residence times of the water in lakes vary greatly. Usually, the average depth of lakes is quite low, except in certain mountainous areas where maximum depths can reach several hundred metres.

Land use and other basin characteristics control the run-off process, so the management of a lake can greatly benefit from the use of geographical information systems (GIS). As the morphological characteristics of the lake itself are of particular importance, a bathymetric map – preferably in a data system format – can be used to define the morphological features, as well as for various physical, chemical and biological studies.

---

[71] UNECE Working Group on Monitoring and Assessment, 2002. *Guidelines for the Monitoring and Assessment of Transboundary and International Lakes. Part A: Strategy document*, available at https://unece.org/DAM/env/water/publications/assessment/lakesstrategydoc.pdf and UNECE Working Group on Monitoring and Assessment, 2003. *Guidelines on Monitoring and Assessment of Transboundary and International Lakes. Part B: Technical guidelines*, available at https://unece.org/DAM/env/water/publications/assessment/lakestechnicaldoc.pdf

The general status of the lake must also be considered. Major discharges must be monitored through a sampling network in order to estimate the effects of loading as a function of distance. Planned sampling of critical ecosystem components that enable assessment of the healthy functioning of the lake ecosystem (e.g. phytoplankton, zooplankton, macrophytes, lake-bottom fauna, fish, invasive species, etc.) is essential to support observations of physical and chemical characteristics in conjunction with simultaneous hydrological observations.

Frequencies

A considerable proportion of hydrological data should be collected in real time or almost real time to allow for efficient lake management. Where data are collected to analyse basic hydrological variability, the requirement for real-time data is not relevant.

In lakes, an important phenomenon is the vertical distribution of temperature in relation to the season. During summertime, thermal stratification can be detected in all deeper lakes. The temperature is highest in the upper water layer, and may match the temperature in rivers. The temperatures in the deeper layer of the lake can remain much colder (5–10 °C) during the entire stratification period. Therefore, in deeper lakes, it is important to consider seasonal vertical temperature distributions in sampling. Many lakes in higher latitudes are dimictic (i.e. the whole water mass only mixes twice a year – in spring and in autumn). Monitoring frequencies should account for such variation.

Locations

The different parts of a lake can have very different characteristics which need to be reflected in the selection of monitoring locations. When planning a sampling network, bathymetric maps and habitat maps should be used in conjunction with suitable information on prevailing currents in the lake. The precise locations of wastewater outlets and other possible sources of pressure must also be known. Sampling sites are usually located in the deepest parts of lakes to allow for the sampling of different layers of water. The number of sampling sites depends on the total area of the lake and the possible existence of separate deeper waters. In addition to sampling deep waters, data will also be needed from lake-bottom areas nearer the shoreline.

Lake Titicaca in Peru

Annex 3. Specific aspects of river monitoring

This annex represents an updated synthesis of information from the *Guidelines on Monitoring and Assessment of Transboundary Rivers.*[72]

Characteristics

Rivers form part of the whole water cycle, and understanding their interactions with other waters is essential to their monitoring. This includes groundwaters and other surface waters (lakes and reservoirs) and the relation between freshwater and receiving coastal and marine waters.

River systems are considered to include tidal estuaries which often have dominating sedimentation problems (polluted sediments, dredging). Given the intense interaction between rivers and the seas into which they discharge, it is essential to harmonize approaches to monitoring and assessment with those adopted under existing sea treaties.

Important variables

To ensure reliable information, it is essential to conduct a systematic analysis and assessment of water quality, flow regimes and water levels, habitats, biological communities, sources and fate of pollutants, as well as mass balance derivations. In addition to water level and river flow, other important aspects of water quantity include sediment discharges, water temperature, evapotranspiration, and ice and snow characteristics.

Flood-risk mapping is a useful management tool to indicate areas that are most vulnerable to flooding in a geographical overview of the river basin. Geo-morphological information on flood plains is needed to estimate the flood frequency of connected areas, and hydrodynamic models can be used to estimate the flood situation in the river during extreme flood events. Model computations should also be used to estimate the impact of human activities on flood risks (e.g. river regulation works, flood protection works and water retention).

The morphology of rivers can alter substantially and have a varying discharge regime, with river dunes formed and subsequently washed away. River training[73] measures are often undertaken to reduce dynamics, but the variation can be huge, especially in larger rivers. This variation may also influence the impacts of flood events, and information on sediment dynamics may therefore be needed.

In addition, low-flow conditions in rivers and drought in the catchment can provoke problems for water uses and the ecological functioning of the river. In the case of drought, more frequent information and data sharing on reservoir operation, diversions and water uses, as well as on hydrological and meteorological parameters, might be necessary.

Regarding river flow conditions, flow management needs to be integrated into overall river management. Application of the concept of environmental flows (e-flows) provides the means for integrated management of river flows to meet the needs of people, agriculture, industry, energy and ecosystems within the limits of available supply and under a changing climate.[74]

Lastly, the assessment of ice conditions on rivers, lakes and reservoirs is of great interest in regions where ice formation affects navigation, interrupts the operation of river-regulating structures, or results in damage to structures, including the formation of ice jams (even to the extent of damming a major river). The obstruction of stream flow by ice can cause serious local flooding.

Frequencies

The continuous or frequent measurement of water levels and river flow is of the utmost importance for river basin management. These basic characteristics play a role in all functions and uses of the river but are especially important for such aspects as water supply, navigation, ecological functions and protection against flooding.

Frequency of measurements, data transmission and forecasting depend on the variability of hydrological characteristics und the response time requirements of the objective of monitoring. The seasonal distribution of flow in rivers depends heavily on their sources. These might include snow or glacier melt, large lakes or base flow

---

[72] UNECE Task Force on Monitoring and Assessment, 2000. *Guidelines on Monitoring and Assessment of Transboundary Rivers*, available at https://unece.org/DAM/env/water/publications/assessment/guidelines_rivers_2000_english.pdf

[73] River training refers to engineering works such as embankments, weirs or dredging to alter the hydrodynamics of the river in support of certain functions, like transport.

[74] See, e.g. IUCN Water Briefing, *Environmental Flows*, available at https://riomagdalenaorg.files.wordpress.com/2019/04/iucn_water_briefing_eflows.pdf

from groundwater, while the beds of rivers flowing into desert sinks may be dry for a significant part of the year. The seasonal distribution of flow also depends on evapotranspiration.

Where streams are concerned, systematic water level recordings, supplemented by more frequent readings during floods, will be needed in the majority of cases. The installation of water level recorders is essential for streams whose level is subject to abrupt fluctuations. Continuous river flow records are necessary to estimate the sediment or chemical loads of streams, including pollutants, and for the design of water-supply systems.

Sampling should be performed preferably at or near border crossings in order to be able to demonstrate the contribution towards reduction targets per country, among other reasons. Sampling in the river and in main tributaries upstream of the confluence is important to show the contribution (e.g. pollution load) of the different tributaries. The selection of sampling sites downstream of a confluence should avoid uncertainties related to incomplete mixing (mixing zones can be several kilometres long, depending on the width-depth ratio of the main river).

In general, locations in the main flow of the river will be chosen for water and suspended solid sampling. Bottom sediment can best be sampled in regions where the suspended material settles. As a consequence, most sediment samples are taken near riverbanks and in the downstream sedimentation area.

The number of sampling points for sediment monitoring depends strongly on the objectives. For trend detection, a low number of sampling points or mixing samples into composite samples can yield enough information. If spatial information is to be estimated, the number of sampling sites will increase, and no composite samples will be used.

Locations

It is necessary to have a clear picture of the part of the river used as a monitoring location for which the monitoring results can be considered representative. A monitoring site can be considered representative in two ways:

- At a macro scale the selection of monitoring sites will be determined by the information objectives (far field representative).
- At a micro scale the local circumstances determine the exact monitoring location (near field representative).

When it is necessary to make combined use of quantity and quality data (e.g. in the calculation of loads), the location of hydrological measurements and water quality sampling should be the same, as far as possible. Different locations are allowed only if the relationship between the hydrological characteristics of both sites is clear and unambiguous.

The selection of monitoring sites for the management of a transboundary river basin should be governed by the purpose for which the data or records are collected and by the accessibility of the site. For hydrometeorological parameters, spatial representativeness is crucial.

Key locations for gauging stations are the lower reaches of rivers, immediately upstream of the river mouth, or where the rivers cross borders near the confluence with tributaries and at major cities along the river (used for flood forecasting, water supply and transport). When evaluating flood risk, relatively high upstream locations or specific tributaries may be important in terms of significant and fast increase in contribution to the flow. In general, a sufficient number of gauging stations should be located along the main river to permit interpolation of water level and discharge between the stations. Water balances will also require a sufficient number of observation stations including at small streams and tributaries.

Monitoring with drones

Annex 4. Specific aspects of monitoring in transitional waters

This annex is based on a few sources[75] as well as the Monitoring Guidelines of the EU Water Framework Directive,[76] which supply specific recommendations for transitional waters.

Characteristics

Transitional waters are water bodies where the rivers meet the sea (i.e. estuaries, deltas, lagoons and coastal brackish water lakes). Transitional waters and their surrounding wetlands are home to unique plant and animal communities that have adapted to brackish water. The mixture of seawater and freshwater in estuaries and in other transitional waters has a salinity in the range from 0.5 to 35 parts per thousand (ppt). The salinity of transitional waters can change from one day to the next depending on the coastal morphology, openness and exposure to the sea, tides, riverine flow, winds and other factors.

The tidal pattern depends on the geographic location, the shape of the coastline and ocean floor, the depth of the water, local winds and any restrictions to water flow. Estuaries in particular are strongly affected by coastal hydrodynamics (e.g. currents, mixing and upwellings), tides and tidal cycles, and the changing freshwater flow from rivers. Many estuaries and particularly coastal lagoons are protected from the full force of ocean waves, winds and storms by reefs, barrier islands, or fingers of land, mud, or sand that surround them.

Important variables

The hydrological budget determines sediment distribution and affects the sensitivity and resilience of transitional water ecosystems. Consequently, the hydrological budget has a major influence on all variables in transitional waters.

Hydrological parameters of relevance for an estuary are the volumes entering the estuary during high and low tide (tidal volume). Both the volume and velocity of the waterflow vary very locally. Subsequent erosion and sedimentation processes are sensitive to anthropogenic measures and extreme events such as storms.

Dissolved oxygen is critical for the survival of animals and plants that live in the water. The temperature of the water also determines the types of plants and animals that can live in the estuary.

Water levels in an estuary typically rise and fall with the daily tides, but periods of drought or excessive rainfall can also affect the amount of freshwater entering the estuary from rivers or runoff, and can easily change the physical, chemical and biological conditions.

Estuarine organisms have different tolerances and responses to salinity changes. The salinity of the water is therefore also an important variable for monitoring. When measuring the salinity gradient, it is essential to do so horizontally as well as vertically.

Turbidity indicates the sediments and other suspended solids in the water. Turbidity affects organisms that are directly dependent on light, such as aquatic plants, because it limits their ability to carry out photosynthesis.

Finally, depending on the goals and objectives of the monitoring, appropriate biological group(s) should be monitored.

Frequencies

Environmental conditions change with the seasons, and monitoring results can reflect those variations. For example, nutrient and pesticide concentrations in estuaries vary considerably from season to season.

Locations

Monitoring locations are selected in accordance with the goals and objectives of the monitoring. Suitable locations will be determined by tidal movements and possible issues that require monitoring. Depending on the morphology of the estuary, stratification may occur. The depth of sampling is therefore also a relevant factor. When monitoring, data collection should be organized with all freshwater inputs and outputs arranged on a seasonal scale.

---

[75]   National Oceanic and Atmospheric Administration website https://oceanservice.noaa.gov/education/tutorial_estuaries/est01_whatis.html and EPA, The Ocean Conservancy, 2006, *Volunteer Estuary Monitoring. A Methods Manual. Second Edition*, available at https://www.epa.gov/sites/default/files/2015-09/documents/2007_04_09_estuaries_monitoruments_manual.pdf

[76]   *Monitoring under the Water Framework Directive*, CIS guidance document No 7, available at https://circabc.europa.eu/sd/a/63f7715f-0f45-4955-b7cb-58ca305e42a8/Guidance%20No%207%20-%20Monitoring%20(WG%202.7).pdf

Pilot monitoring exercise Armenia-Georgia, September 2020

Annex 5. International programmes and information sources

Environmental data and information are available through various online databases and websites. Monitoring and assessment activities under the auspices of United Nations organizations and programmes produce valuable information which can be used for carrying out assessments of transboundary waters. This annex provides a non-comprehensive selection of such programmes and information sources.

The **GEMS/Water Programme**[77] is a major source of global surface water and groundwater quality data and provides information on the state and trends of regional and global water quality to support scientific assessments and decision-making.

The **Transboundary Waters Assessment Programme (GEF TWAP)**[78] contains information on biophysical parameters of waterbodies (including biodiversity, climate change, land degradation, waste, etc.), socio-economic information (including population, Human Development Index [as a proxy for consumption], etc.) and governance-related parameters (including multilateral environmental arrangements, integrated national planning, etc.).

AQUASTAT[79] is the FAO global information system on water resources and agricultural water management. It collects, analyses and provides free access to over 180 variables and indicators by country dating back to 1960. Data and metadata are available on water resources (internal, transboundary and total), water uses (by sector and source), wastewater, irrigation (location, area, typology, technology and crops), dams (location, height, capacity and surface area), and water-related institutions, policies and legislation. FAO has also developed **WaPOR**,[80] a publicly accessible near real-time database that uses satellite data to enable monitoring of agricultural water productivity. WaPOR data also helps countries monitor agricultural water productivity, identify water productivity gaps and find solutions.

Data and information on groundwaters can be obtained from the **International Shared Aquifer Resources Management (ISARM)** programme,[81] which develops methods and techniques aimed at improving understanding of the management of shared groundwater systems, and considers both technical and institutional aspects. The **International Groundwater Resources Assessment Centre (IGRAC)**[82] facilitates and promotes worldwide exchange of groundwater knowledge to improve assessment, development and management of groundwater resources. The centre's **Global Groundwater Information System**[83] is an interactive, web-based portal that provides access to groundwater-related information and knowledge.

The national hydrological/meteorological services of the member states of the World Meteorological Organization (WMO) operate over 475 000 hydrological stations worldwide. WMO's **Associated Programme on Flood Management (APFM)**[84] aims to support countries in implementing Integrated Flood Management to maximize net benefits from floodplains and minimize loss of life from flooding. The organization's **Integrated Drought Management Programme (IDMP)**[85] provides policy and management guidance through globally coordinated generation of scientific information and sharing of best practices and knowledge for integrated drought management. WMO's **Global Runoff Data Centre (GRDC)**[86] is a digital worldwide depository of discharge data and associated metadata and serves as a facilitator between data providers and data users.

Data on water-related disease can be accessed through the **Health for All Database**[87] of the World Health Organization (WHO). This database includes data on diarrhoeal diseases, viral hepatitis A and malaria incidence, as well as information on the number of people connected to water supply systems and with access to sewage systems, septic tanks or other hygienic sewage disposals. The WHO/UNICEF **Joint Monitoring Programme for Water Supply, Sanitation and Hygiene (JMP)** database[88] provides global data on water supply, sanitation and hygiene.

---

[77] www.gemswater.org

[78] www.geftwap.org

[79] www.fao.org/aquastat/en

[80] https://wapor.apps.fao.org/home/WAPOR/1

[81] https://isarm.org

[82] www.un-igrac.org

[83] https://ggis.un-igrac.org

[84] www.floodmanagement.info

[85] www.droughtmanagement.info

[86] www.bafg.de/GRDC/EN/Home/homepage_node.html

[87] www.euro.who.int/hfadb

[88] https://washdata.org/data

The **SDG Global Database**[89] provides data on more than 210 SDG indicators for countries across the globe. UN-Water supports countries in monitoring water- and sanitation-related indicators of SDG 6.[90]

Google Earth Engine[91] provides a wide range of geophysical and weather and climate datasets.

At the regional level, one important source of information on the status of rivers, lakes and groundwaters in Europe is the **European Environment Agency (EEA)**.[92] **Copernicus**[93] is the European Union's Earth observation programme, offering information services that draw from satellite Earth Observation and in-situ (non-space) data. Copernicus provides services related to the atmosphere, the marine environment, land cover and land use, climate change, border security and early warning. The **Statistical Office of the European Communities (Eurostat)**[94] collects statistics on water resources, water abstraction and use, and wastewater treatment and discharges.

In Asia, the member states of the Association of Southeast Asian Nations have established the **ASEAN Hydroinformatics Data Centre (AHC)**[95] **for Water and Disaster Risk Management** in effort to address climate change.

The Regional Environmental Centre for Central Asia (CAREC) runs the **Eurasian River Basin Portal**,[96] which provides support to water resources management and aims to strengthen the capacity of water management organizations in Europe and Central Asia, and functions as an information portal on climate adaptation and mitigation in Central Asia.[97]

The **Central Asia Regional Water Information Base Project (CAREWIB)**[98] implemented by SIC ICWC provides information on water and environment in Central Asian countries.

Test at a borehole, Senegal

---

[89] https://unstats.un.org/sdgs/unsdg

[90] www.sdg6monitoring.org and www.sdg6data.org

[91] https://earthengine.google.com

[92] www.eea.europa.eu/data-and-maps

[93] www.copernicus.eu/en

[94] https://ec.europa.eu/eurostat/web/environment/water

[95] www.aseanwater.net/wp

[96] www.riverbp.net/eng

[97] https://ca-climate.org/eng

[98] http://www.cawater-info.net/about_e.htm

Sewage treatment plant, Koblenz, Germany